The Drones New Era

Education, Volume 4

Artur Victoria

Published by Artur Victoria, 2025.

THE DRONES NEW ERA

First edition. June 13, 2025.

ISBN: 979-8231630530

Written by Artur Victoria.

Introduction

Drones Evolution

Strategic and Tactical Advantages

The US War to terrorism - Afghanistan, Pakistan, and Yemen

The Israeli use of drones in the Gaza Strip

Drones and the Ukrainian War

Drone Warfare in the Middle East

The Ethical and Legal Dimensions of Drone Warfare

The Lack of International Laws and Regulation

The Rise of Military Drones

The Future of Drones

The Future of Military Strategy and International Security

Artificial Intelligence (AI) Integration

Future of Autonomous Drone Technology

Hypersonic and High-Speed Drones The Potential of Drone Technology

Integration Technologies – Air – Land and Sea

Introduction

Unmanned aerial vehicles (UAVs), commonly referred to as drones, have irrevocably altered the scene of modern warfare. Their proliferation across the globe, employed by both state and non-state actors, necessitates a thorough understanding of their capabilities, limitations, and implications for military strategy and international security. This book serves as a foundational introduction to the world of drone warfare, providing a comprehensive overview of the technology, its applications, future technologies development and its far-reaching consequences.

At its core, a drone is an aircraft without a human pilot onboard. This seemingly simple definition encompasses a vast assortment of designs, capabilities, and levels of autonomy. The term "drone" it is a colloquialism; the more precise and widely accepted term is Unmanned Aerial Vehicle (UAV), a designation reflecting the absence of a human pilot and the vehicle's aerial nature. However, the term "drone" has become so entrenched in popular usage that both terms are employed interchangeably throughout this text. The spectrum of UAVs is astonishingly broad, ranging from small, hand-launched devices weighing only a few pounds to massive, high-altitude platforms capable of sustained flight for days. In the future nano drones will be practically invisible.

One of the most critical distinctions among UAVs lies in their airframe design. Fixed-wing drones, resembling miniature airplanes, are optimized for speed and long-range flight. Their aerodynamic efficiency allows them to cover vast distances, making them ideal for surveillance and reconnaissance missions over extended periods. On the other hand, rotary-wing drones, commonly known as quadcopters or multirotors, offer superior manoeuvrability and hovering capabilities. Their vertical takeoff and landing (VTOL) capability makes them well-suited for operations in confined spaces and challenging terrains. Hybrid designs, combining fixed-wing and rotary-wing elements, are emerging, aiming to harness the strengths of both architectures, offering extended range with enhanced agility.

The propulsion system is another key differentiator. Early UAVs relied heavily on internal combustion engines, similar to those found in model airplanes. These systems, however, are noisy, produce significant emissions, and possess inherent limitations in terms of efficiency and endurance. More recent advancements have led to a widespread adoption of electric motors, particularly in smaller drones. Electric propulsion offers quieter operation, reduced emissions, and increased efficiency. The limitations of battery technology, however, often restrict flight times. Hybrid propulsion systems, combining internal combustion engines with electric motors, are increasingly prevalent, seeking to maximize both range and endurance. Fuel cell technology is also showing promise, potentially offering significantly extended flight times in the future.

The payload capacity of a UAV determines its operational capabilities. Small drones might carry a small camera and communication equipment and nano weapons, while larger platforms can accommodate sophisticated sensor suites, electronic warfare systems, and even guided munitions. This variation in payload is a critical factor in determining the specific military applications of different drone models.

The sensor suite is arguably the most critical aspect of a UAV's capabilities. Depending on the mission parameters, drones can be equipped with an array of sensors to gather valuable intelligence. Electro-optical (EO) cameras provide high-resolution imagery in visible light, while infrared (IR) cameras detect thermal signatures, allowing for target identification even in low-light conditions or through obscurants like smoke or fog. Synthetic Aperture Radar (SAR) provides high-resolution images regardless of weather or lighting conditions. Light Detection and Ranging (LiDAR) systems create detailed three-dimensional maps of the terrain, invaluable for navigation and target acquisition. Electronic intelligence (ELINT) gathering systems can intercept and analyze radio frequency signals, providing critical information about enemy communications and electronic systems. The combination and sophistication of these sensors significantly influence a drone's effectiveness in reconnaissance, surveillance, and target acquisition.

Reliable communication links are indispensable for effective drone operation. Line-of-sight (LOS) communication relies on a direct path between the drone and the ground control station, limiting operational range. Beyond-line-of-sight (BLOS) communication, often facilitated by satellite links, extends the operational range considerably, allowing for longer missions and broader area coverage. The reliability and security of these communication links are of vital importance, as disruptions could lead to mission failure or even loss of the drone. Advanced data encryption and other cyber security measures are critical to protecting drone communications from interception or jamming by adversaries.

The level of autonomy is another defining characteristic of UAVs. Remotely piloted vehicles (RPVs) are controlled directly by a human operator, providing a degree of fine-grained control over the drone's actions. Autonomous systems, on the other hand, possess a degree of self-decision-making capability, enabling them to execute complex tasks without constant human intervention. This autonomy can range from simple waypoint navigation to complex decision-making algorithms for target selection and engagement. The level of autonomy is a critical consideration, balancing the advantages of reduced operator workload with the risks associated with delegating potentially lethal decisions to machines.

The historical development of drone technology reveals a fascinating progression from rudimentary radio-controlled aircraft to the highly sophisticated platforms deployed today. The early days saw the development of simple target drones used for anti-aircraft gunnery training. Over time, technological advancements in areas like avionics, propulsion, and sensor technology led to increasingly sophisticated platforms, capable of performing increasingly complex missions. The development of robust satellite communication systems played a fundamental role in extending the operational range of UAVs, paving the way for their use in diverse military contexts.

The global scenery of drone manufacturers and military users is highly dynamic and constantly evolving. A handful of major players dominate the market for larger, military-grade drones, but a significant number of smaller companies produce more affordable, commercially available drones, contributing to their wider availability across the globe. The widespread adoption of drone technology by a growing number of military forces reflects the undeniable strategic advantages they offer. The balance of power in conflicts can be heavily influenced by the proliferation and sophistication of drone technology in the hands of various nations.

Specific examples of drone models and their capabilities further illustrate this diversity. The General Atomics MQ-9 Reaper, a large, long-endurance UAV, is renowned for its extensive surveillance capabilities and its capacity to carry a significant payload of precision-guided munitions. The smaller Boeing Insitu ScanEagle is designed for reconnaissance and surveillance missions, highlighting the vast difference in mission roles and capacities achievable by different platforms. The Chinese CH-4B Rainbow drone, commercially available, has been deployed by various non-state actors, highlighting the concerning accessibility of certain models. Each model illustrates a specific niche, driven by specific operational needs and technological capabilities. Understanding this diverse scenery is essential to grasping the full strategic implications of drone warfare.

Their rapid evolution is driven by a confluence of factors, including advancements in sensor technology, autonomous navigation systems, artificial intelligence (AI), propulsion systems, and materials science. Understanding these technological leaps is essential to comprehending the evolving nature of drone warfare and its implications for military strategy.

One of the most significant advancements lies in the sophistication of sensor systems. Early drones often relied on relatively rudimentary cameras providing limited imagery. Today, however, UAVs can be equipped with an array of advanced sensors capable of gathering a wealth of information. High-resolution electro-optical (EO) cameras provide incredibly detailed imagery, allowing for the identification of even small targets. Thermal imaging systems detect heat signatures, enabling target detection regardless of ambient light conditions or camouflage. This is especially valuable in night-time operations or when targets are obscured by foliage or other obstacles.

Synthetic Aperture Radar (SAR) technology represents a significant leap forward in all-weather intelligence gathering. Unlike EO cameras, SAR systems can penetrate clouds, fog, and even some types of foliage, producing high-resolution images regardless of weather conditions. This capability is invaluable in environments with persistent cloud cover or adverse weather, providing a constant stream of intelligence even when other sensor systems are compromised. Furthermore, SAR can be used to create detailed maps of the terrain, assisting with navigation and target identification.

Light Detection and Ranging (LiDAR) technology complements SAR by creating highly accurate three-dimensional models of the terrain. By emitting laser pulses and measuring the time it takes for the light to reflect back, LiDAR systems can generate detailed point cloud data, creating precise representations of the environment, including buildings, trees, and other obstacles. This capability is particularly useful in urban environments or complex terrain where accurate navigation and precise target designation are paramount.

Beyond visual and terrain mapping sensors, UAVs are increasingly incorporating electronic warfare (EW) capabilities. These systems enable drones to detect, analyze, and even jam enemy radar and communications signals. This ability to disrupt enemy communication networks and blind their sensors provides a significant tactical advantage, hindering their ability to coordinate and respond effectively. The integration of EW capabilities into drones makes them a powerful force multiplier, enabling them to operate effectively even in contested electromagnetic environments.

Drones Evolution

The development of sophisticated autonomous navigation systems is a critical aspect of the ongoing technological revolution in drone capabilities. Early drones relied heavily on direct control by a human operator, often requiring a clear line of sight. Modern drones, however, utilize advanced GPS and inertial navigation systems, allowing them to navigate autonomously over long distances and in challenging environments. This increased autonomy reduces the workload on human operators, enabling them to control multiple drones simultaneously. Furthermore, advanced algorithms allow drones to adapt to changing conditions, rerouting around obstacles or adjusting flight plans based on real-time data.

The integration of artificial intelligence (AI) is rapidly transforming drone capabilities. AI algorithms are used for various tasks, including autonomous target recognition, mission planning, and route optimization. AI-powered systems can analyze vast quantities of sensor data in real-time, identifying potential threats and guiding drones toward their objectives. The use of AI in drone operation also improves situational awareness, providing human operators with central information that may otherwise be missed. The potential for AI to enhance the effectiveness of drones is immense, but it also raises ethical considerations related to autonomous targeting and decision-making.

Advancements in propulsion systems are also contributing to the enhanced performance of UAVs. Early drones relied on combustion engines, which are relatively noisy and produce considerable emissions. More recent designs often utilize electric motors, offering quieter operation and reduced environmental impact. The development of high-energy-density batteries is significantly extending flight times, allowing drones to remain airborne for extended periods. Hybrid propulsion systems, combining internal combustion engines with electric motors, offer a balance between power and endurance, providing optimal performance for specific missions. Fuel cell technology is also emerging as a promising alternative, potentially offering significantly extended flight times in the future.

Materials science is also playing a critical role in advancing UAV capabilities. The use of lightweight yet high-strength composite materials allows drones to be more manoeuvrable and less vulnerable to damage. These advanced materials also contribute to reduced weight, extending flight times and increasing payload capacity. The development of stealth technologies, utilizing radar-absorbing materials, makes drones more difficult to detect by enemy radar systems, enhancing survivability in contested airspace.

The integration of sophisticated data analytics techniques is revolutionizing the efficiency of drone operations. Advanced algorithms process sensor data from multiple sources, providing a comprehensive picture of the operational environment. This data fusion enhances situational awareness, enabling better decision-making during missions. Post-mission analysis of the collected data provides valuable insights that can be used to improve future operations, refining mission planning and optimizing drone performance. These analytical tools are essential for fully leveraging the potential of drone technology, extracting maximum value from the data collected.

The use of high-resolution cameras on drones has significantly improved reconnaissance capabilities, enabling military forces to monitor enemy movements and activities with unprecedented detail. The development of autonomous navigation systems has expanded the range of potential drone deployments, enabling missions in remote and hazardous environments. The integration of AI has enabled drones to execute increasingly complex tasks with a high degree of autonomy, reducing the risk to human pilots and maximizing mission effectiveness. The use of swarm technology, where multiple drones coordinate their actions to achieve a common objective, is becoming increasingly prevalent, demonstrating a new level of tactical capability.

However, the technological advancements in drone capabilities also present significant challenges. The increased autonomy of drones raises ethical concerns about the potential for unintended consequences, particularly in regard to civilian casualties. The proliferation of drone technology raises concerns about its potential misuse by non-state actors. The development of counter-drone technologies is central to mitigate the risks associated with the growing use of drones in military operations.

Moving up in scale, medium-altitude long-endurance (MALE) drones represent a significant advancement in unmanned aerial vehicle technology. These systems, characterized by their extended flight times and greater payload capacities, are capable of undertaking sustained surveillance missions, precision strikes, and intelligence gathering operations over vast geographical areas. The larger size allows for the integration of more sophisticated sensor packages, including high-resolution EO/IR cameras, SAR, and signals intelligence (SIGINT) payloads. MALE drones often serve as persistent surveillance platforms, providing a continuous stream of intelligence data to commanders. Their endurance allows them to remain airborne for days, providing continuous monitoring of large areas, an ability that significantly enhances strategic planning and tactical decision-making.

The MQ-9 Reaper is a prime example of a successful MALE drone, utilized extensively by NATO airforces. Its ability to carry a substantial payload, including precision-guided munitions, enables it to carry out both surveillance and strike missions. Its operational history showcases the effectiveness of MALE drones in counterterrorism operations, intelligence gathering, and battlefield support. However, MALE drones are vulnerable to sophisticated air defence systems and sophisticated electronic warfare tactics. Their larger size and longer flight times make them more easily detected and targeted, emphasizing the importance of defensive countermeasures and strategic deployment to mitigate such risks. The logistical support required for these systems is also considerably more complex than that needed for smaller tactical drones, including infrastructure for launching, recovering, and maintaining these systems.

At the top of the UAV hierarchy are the large, high-altitude drones designed for strategic intelligence gathering and long-range surveillance. These systems operate at altitudes exceeding those of most manned aircraft, providing an unparalleled vantage point for observing large swathes of territory. Their extended range and endurance capabilities allow them to monitor vast areas for prolonged periods, even crossing international boundaries. These high-altitude platforms typically employ advanced sensor packages, including advanced SIGINT capabilities and sophisticated image processing systems to analyze gathered data. The data collected is vital for strategic planning, early warning systems, and tracking enemy movements across larger theatres of operation.

Examples of high-altitude drones include the RQ-4 Global Hawk, deployed by the US Air Force for wide-area surveillance and reconnaissance. Its high-altitude operation allows it to avoid many threats present at lower altitudes. The high-resolution imagery and SIGINT capabilities are integral to military intelligence, allowing analysts to monitor movements of enemy forces, infrastructure development, and other key indicators of intent. However, the high cost and specialized infrastructure required to operate these systems limit their widespread adoption. Maintenance and operational costs are substantial, which often restricts their usage to larger military forces. Additionally, their vulnerability to sophisticated anti-aircraft weapons and jamming technologies highlights the ongoing arms race between drone technology and counter-drone measures.

The integration of drones across different branches of the military further underscores their adaptability and increasing importance. The Army often employs smaller, tactical drones for close-range reconnaissance, battlefield assessment, and providing immediate situational awareness to ground troops. The Navy use both smaller and larger drones for maritime surveillance, anti-submarine warfare, and strike capabilities, extending their operational reach beyond the horizon. Meanwhile, the Air Force utilizes larger MALE and high-altitude drones for long-range surveillance, strike operations, and gathering strategic intelligence over broad geographical regions. The specialized requirements of each branch influence the design, capabilities, and operational roles of drones used.

Beyond the core capabilities of surveillance and strike, UAVs are increasingly equipped with diverse payloads. Communication relays mounted on drones extend communication ranges in remote or contested environments. These relay systems provide critical links between ground troops, airborne assets, and command centres, improving communication efficiency and situational awareness. The integration of electronic warfare systems into drones expands their tactical capabilities, allowing them to disrupt enemy communications and sensors, gaining an electronic warfare advantage. This ability to deny the enemy critical information and interfere with their operations is proving increasingly significant in modern conflicts. The development and integration of increasingly sophisticated payloads will further expand the operational versatility of drones in the years to come.

Strategic and Tactical Advantages

The strategic and tactical advantages of drones are undeniable: increased persistence, improved situational awareness, precise targeting capabilities, and reduced risk to human life. However, limitations exist. The reliance on satellite communication links creates vulnerabilities in areas with limited satellite coverage or during periods of electronic jamming. Their vulnerability to sophisticated air defence systems and cyber attacks presents an ongoing challenge, necessitating robust security measures and counter-measures.

As drone technology continues to evolve, addressing these challenges will be critical to ensuring its responsible and effective utilization. The future of military operations will undoubtedly be shaped by the continued advancements and integration of UAVs, requiring a constant reassessment of operational strategies and ethical implications.

The proliferation of unmanned aerial vehicles (UAVs) has fundamentally reshaped the scene of modern warfare, impacting both strategic and tactical military operations in profound ways. Their integration into military doctrine has led to a revolution in intelligence, surveillance, and reconnaissance (ISR), fundamentally changing how conflicts are planned, executed, and ultimately won. The ability to deploy persistent surveillance platforms, capable of gathering real-time intelligence across vast geographical areas, has provided military commanders with an unprecedented level of situational awareness. This enhanced awareness enables more effective targeting, reducing collateral damage and improving the overall efficiency of military operations.

Prior to the widespread adoption of drones, intelligence gathering relied heavily on human assets, manned reconnaissance aircraft, and satellite imagery. These methods were often time-consuming, expensive, and risked the lives of personnel. Drones, however, offer a cost-effective and relatively low-risk alternative, providing a continuous stream of real-time intelligence without jeopardizing human lives. The persistent surveillance capabilities of drones like the MQ-9 Reaper, for example, have allowed for the prolonged monitoring of suspected terrorist camps, enabling the identification of high-value targets and the planning of precise, surgical strikes.

The impact on tactical operations is equally significant. Drones have proven invaluable in close-quarters combat situations, offering ground troops real-time imagery from overhead perspectives. Small, tactical drones, easily deployed and operated, provide immediate situational awareness, enabling commanders to make informed decisions in dynamic battlefield conditions. These small UAVs are particularly useful in urban environments or difficult terrains where traditional reconnaissance methods might be impractical or too dangerous. Their small size and manoeuvrability also make them harder to detect and engage, increasing their survivability.

The ability of drones to deliver precision strikes has also changed the dynamics of modern warfare. The use of precision-guided munitions (PGMs) delivered by drones allows for surgical strikes against specific targets, minimizing collateral damage and civilian casualties. This capability is especially important in counterterrorism operations, where the goal is to neutralize enemy combatants while protecting innocent civilians. The inherent precision of drone strikes, when compared to traditional air strikes, reduces the likelihood of accidental deaths and collateral damage, enhancing the legitimacy and ethical implications of military actions. However, this precision is not without its limitations; the accuracy is still dependent upon accurate targeting intelligence and the complex nature of target identification.

The ease with which drones can be deployed and the relative lack of risk to the operators have raised concerns about the potential for increased reliance on lethal force. The removal of physical proximity between the operator and the target can potentially desensitize operators to the consequences of their actions, leading to an increased risk of civilian casualties and potentially sparking international controversy. This ethical dilemma requires constant oversight, stricter protocols and improved target identification methods and techniques.

The use of drones in counterterrorism operations has been a focal point of debate. On the one hand, drones have allowed for the targeted elimination of high-value terrorist leaders, disrupting terrorist networks and preventing potential attacks. On the other hand, the use of drones has also raised concerns about potential civilian casualties and the potential for unintended consequences. The secrecy surrounding many drone strikes and the lack of transparency have motivated criticism and mistrust, particularly among affected populations. The need for a greater level of accountability and transparency in drone operations is vital to address these ethical concerns.

Beyond the realm of counterterrorism, drones are also playing an increasingly important role in other military operations. Their use in maritime surveillance, border security, and disaster relief operations highlights their versatility and adaptability. The ability of drones to provide persistent surveillance over large areas makes them invaluable assets for monitoring maritime traffic, detecting illicit activities, and securing borders. Their use in disaster relief operations allows for a rapid assessment of damage, delivery of aid, and coordination of rescue efforts. The inherent advantages of drone technology for both military and civilian applications are being actively explored.

However, the strategic and tactical advantages of drone warfare are not without limitations. Drones are vulnerable to electronic warfare (EW), sophisticated air defence systems, and cyber attacks. The reliance on satellite communication links can be a significant vulnerability in areas with limited satellite coverage or during periods of electronic jamming. The susceptibility of drones to sophisticated EW can disrupt their operations, rendering them ineffective or even turning them against their operators. This reality underscores the ongoing arms race between drone technology and counter-drone measures.

Furthermore, the proliferation of drone technology raises concerns about the potential for escalation and destabilization. The relatively low cost and ease of acquisition of drones make them accessible to a wide range of actors, including state and non-state actors. This proliferation has led to an increase in the number of conflicts involving drones, raising concerns about the potential for unintended escalation and accidental warfare. The increased reliance on autonomous systems raises even more serious implications, especially regarding the ethical concerns surrounding lethal autonomous weapons systems (LAWS). The lack of human intervention in the decision-making process within LAWS increases the risk of errors and exacerbates the ethical considerations surrounding the use of lethal force.

The increasing integration of artificial intelligence (AI) into drone technology presents both opportunities and challenges. While AI enhances capabilities such as automated target recognition and autonomous navigation, it also raises concerns about the potential for unintended consequences and bias in algorithmic decision-making. The potential for errors and unintended outcomes underscores the importance of continued research and development to ensure the safety and ethical use of AI in drone technology.

The proliferation of unmanned aerial vehicles (UAVs), or drones, has fundamentally reshaped the conduct of modern warfare, moving beyond the theoretical discussions into the realm of tangible, real-world impact. To fully grasp the implications of this technological shift, it is essential to analyze specific instances where drones have been deployed, examining their effectiveness, limitations, and the often-unintended consequences that accompany their use. The following case studies explore the diverse applications and impacts of drone warfare across various geopolitical contexts.

The US War to terrorism - Afghanistan, Pakistan, and Yemen

One of the most extensively documented examples of drone warfare is the United States' drone program in Afghanistan, Pakistan, and Yemen. Initiated following the 9/11 attacks as part of the broader "War on Terror," the program involved the targeted killing of suspected al-Qaeda and other militant group members. The MQ-1 Predator and MQ-9 Reaper drones became the workhorses of this program, their persistent surveillance capabilities and precision-strike potential providing a new dimension to counterterrorism operations. The program's proponents highlighted its effectiveness in eliminating high-value targets, disrupting terrorist networks, and reducing the risk to American ground troops. The argument was that drones offered a surgically precise tool that minimized civilian casualties when compared to traditional air strikes.

However, the program also generated considerable controversy. Reports of civilian casualties, although often disputed, cast a significant shadow on the program's supposed precision. The lack of transparency surrounding the targeting process, the criteria used to select targets, and the overall number of strikes stimulated criticism and suspicion. Human rights organizations and international bodies raised serious concerns about the legal and ethical implications of targeted killings conducted outside established legal frameworks, highlighting the difficulties in ensuring accountability and due process. The long-term consequences of the program, including the fostering of anti-American sentiment and the potential for radicalization, remain subjects of ongoing debate and analysis. The program's impact on the ground, specifically the disruption of tribal structures and societal norms in targeted areas, requires further extensive study to understand the full implications of this form of asymmetric warfare.

The Israeli use of drones in the Gaza Strip

The Israeli use of drones in the Gaza Strip provides another compelling case study. Facing persistent rocket attacks from Hamas and other Palestinian militant groups, Israel has employed drones extensively for both surveillance and offensive operations. The smaller, tactical drones played an important role in gathering intelligence, while larger UAVs delivered precision strikes against suspected militant targets.

Similar to the US experience, the Israeli drone program has been effective in disrupting militant activities and reducing the overall threat posed by Palestinian groups. However, the program has also been associated with civilian casualties, powering accusations of disproportionate force and violations of international humanitarian law.

The highly concentrated population density of the Gaza Strip exacerbates the challenges in distinguishing between combatants and civilians, underscoring the limitations of even the most advanced targeting systems. Furthermore, the psychological impact of persistent drone surveillance on the civilian population must be considered a significant factor when assessing the overall consequences of Israeli drone operations in Gaza.

The use of drones has not been limited to state actors; non-state actors have increasingly adopted UAV technology, presenting new challenges to national security. Groups like ISIS demonstrated their ability to acquire and deploy drones for surveillance, reconnaissance, and even limited offensive operations. The relatively low cost and ease of acquisition of commercially available drones, coupled with the availability of open-source information on drone modification and weaponization, presents a significant security challenge. The rise of drone technology in non-state actors' hands not only increases the potential for attacks, but also raises concerns about potential escalation and the destabilization of volatile regions. The ease of access to this technology requires new forms of counter-drone measures and international cooperation to curtail the uncontrolled proliferation of armed drones outside state control. The impact on ground operations needs to address the tactical advantages gained by non-state actors utilizing these relatively accessible technologies.

Drones and the Ukrainian War

The Ukrainian conflict has served as a proving ground for the Turkish-made Bayraktar TB2 drone, catapulting it from a relatively unknown military asset to a global symbol of asymmetric warfare. Its effectiveness against Russian armoured columns, particularly in the initial stages of the invasion, surprised many military analysts and significantly altered the battlefield dynamics. The TB2's success wasn't merely due to its technological capabilities but also stemmed from a confluence of factors, including Ukraine's adept integration of the system, the shortcomings in Russian electronic warfare capabilities, and the overall operational context of the war.

The Bayraktar TB2 is a medium-altitude, long-endurance (MALE) drone, possessing a relatively long flight time and substantial payload capacity. Its operational range allows it to cover significant swathes of territory, making it ideal for surveillance, reconnaissance, and precision strikes. Armed with laser-guided munitions, the TB2 can effectively target tanks, armoured personnel carriers, artillery pieces, and other high-value military assets. The drone's relatively low cost compared to manned aircraft offers a significant advantage, allowing Ukraine to deploy multiple systems simultaneously, overwhelming Russian air defences and spreading the risk.

Early footage released on social media and mainstream news channels showcased the TB2's devastating effect on Russian armoured convoys. Videos of TB2 strikes, often shared widely, depicted the destruction of numerous tanks, armoured vehicles, and military equipment, effectively highlighting the drone's battlefield prowess. These visual demonstrations of success not only boosted Ukrainian morale but also had a considerable psychological impact on Russian forces, demonstrating their vulnerability to a relatively inexpensive drone system. The precision targeting capabilities of the TB2, minimizing civilian casualties relative to conventional bombing, also added to its strategic value. This precision significantly countered Russian claims of indiscriminate attacks by Ukrainian forces, a crucial element in the information war surrounding the conflict.

However, the TB2's success was not solely dependent on the drone's inherent capabilities. Ukraine's skilled operators played a vital role in its effectiveness. The integration of the TB2 into existing command-and-control structures and its effective coordination with ground forces highlight the importance of operator training and logistical support in achieving operational success. The use of readily available commercial satellite imagery, coupled with ground-based intelligence gathering, facilitated accurate targeting and efficient mission planning.

Furthermore, the Russian military's initial response to the Bayraktar threat proved inadequate. Reports indicate that the effectiveness of Russian electronic warfare systems in countering the TB2 was limited, particularly in the early phases of the conflict. This lack of effective countermeasures allowed the drones to operate with relative impunity, further enhancing their tactical advantage. The Russian army’s outdated air defence systems, coupled with insufficient coordination among different branches of their armed forces, exacerbated their vulnerability to the TB2. The initial failure to effectively integrate reconnaissance capabilities with ground forces highlights a lack of preparedness for the kind of asymmetric warfare tactics employed by Ukraine.

The strategic implications of the TB2's deployment extend beyond its immediate tactical successes. The drone's impact on the war's information landscape should not be overlooked. The widespread dissemination of video evidence of successful strikes motivated both Ukrainian morale and international support for the war effort. The visual nature of the conflict, widely disseminated through social media, undercut the Russian narrative of swift victory and revealed the realities of the fighting on the ground. This visual evidence played a critical role in shaping international opinion and influenced the flow of foreign aid to Ukraine.

This success story, however, is not without its complexities and caveats. As the war progressed, Russia adapted its tactics to counter the TB2's effectiveness. Reports suggest increased deployment of electronic warfare systems, improved air defence networks, and more effective coordination among units. The use of more sophisticated air defence systems and jamming technologies significantly reduced the operational effectiveness of the TB2 in later stages of the conflict. Furthermore, the relatively low cost and ease of use of the TB2 have also led to concerns regarding proliferation. The risk of the drone falling into the wrong hands, including those of non-state actors or hostile governments, presents a potential long-term security concern.

The operational limitations of the TB2 must also be addressed. While effective against certain targets, the drone's limited payload capacity and vulnerability to sophisticated air defence systems restrict its overall application. Its success heavily depends on favourable weather conditions, terrain, and the absence of advanced electronic countermeasures. These limitations expose a need for continuous technological upgrading and adaptation in the face of evolving warfare strategies. The vulnerability of the TB2's command and control infrastructure also presents an ongoing challenge. Disrupting this network would significantly impact its operational capabilities, highlighting the need for redundant and secure communication systems.

The experience with the Bayraktar TB2 in the Ukraine war has presented a significant paradigm shift in military thinking, particularly in relation to the effectiveness of asymmetric warfare techniques. It has demonstrated the potential of affordable, relatively simple drone technology to significantly alter the balance of power on the battlefield. Its impact, however, is not only tactical; it's strategic, influencing the course of the war, shaping international perceptions, and highlighting the critical role of effective military integration and the adaptability of a modern military to asymmetric threats. The lessons learned from the Ukrainian experience offer valuable insights for military planners and policymakers across the globe. The war's early stages underscored a need for robust electronic warfare capabilities and effective air defence systems, but also demonstrated the effectiveness of a well-integrated, adaptable, and resourceful opponent equipped with simpler technology when employed intelligently. The impact of the Bayraktar TB2 on the Ukraine conflict extends far beyond the battlefield; it has prompted a reassessment of military doctrines, technological investments, and the evolving nature of warfare in the 21st century. The cost-effectiveness and the significant impact of the Bayraktar TB2 have spurred a global reconsideration of drone technology's role in modern warfare, impacting defence budgets and acquisition strategies worldwide. The narrative surrounding the TB2 has permeated popular culture and political discourse, making it a symbol of both technological advancement and the challenges of asymmetric warfare in the modern era. Understanding its successes and limitations offers critical insights into the evolving dynamics of modern warfare and the vital need for adaptable strategies to address the complex challenges of future conflicts.

The narrative surrounding the Bayraktar TB2, while compelling, represents only one facet of the drone warfare revolution reshaping the Ukrainian battlefield. A contrasting, and arguably equally significant, element is the deployment of Iranian-made Shahed drones by Russia. Unlike the Bayraktar TB2's emphasis on precision strikes and relatively high survivability, the Shahed series, particularly the Shahed-136, showcases a different approach to drone warfare: a strategy of attrition through sheer numbers and relative expendability. This divergence in design philosophy reflects fundamentally different operational doctrines and strategic objectives.

The Shahed-136, a loitering monition, or "kamikaze drone," operates on a fundamentally different principle than the Bayraktar TB2. Instead of attempting repeated sorties with a higher degree of precision, the Shahed-136 is designed for a single, one-way mission. Its relatively low cost and simple design enables mass production and deployment, allowing Russia to saturate Ukrainian air defences with a swarm of drones, overwhelming their capabilities through sheer volume. While less sophisticated in terms of guidance and payload than the TB2, the Shahed-136's effectiveness lies in its ability to inflict damage through sheer numbers and the psychological effect of constant attacks.

The relatively simple guidance system of the Shahed-136, reliant on GPS and inertial navigation, makes it vulnerable to electronic countermeasures and susceptible to GPS jamming. However, its low cost makes the loss of individual drones acceptable, a cost-benefit analysis that contrasts sharply with the more expensive and sophisticated Bayraktar TB2. The inherent limitations in precision targeting are compensated for by the volume of drones deployed, increasing the probability of achieving a desired level of damage, even if individual strikes are less precise.

Reports from Ukraine indicate that the Shahed drones have targeted critical infrastructure, including power grids, energy facilities, and water treatment plants. Attacks on civilian infrastructure represent a significant departure from the more targeted military strikes seen with the Bayraktar TB2. This targeting strategy reflects Russia's apparent aim to cripple Ukraine's civilian infrastructure and to undermine public morale through prolonged disruption of essential services. The devastating impact of these attacks on Ukrainian civilians and their infrastructure underscores the ethical and humanitarian considerations raised by the use of drones in conflict, highlighting the differing operational philosophies between the two sides.

The Ukrainian military has responded to the Shahed drone threat with a multi-pronged approach. Initial reports suggested a relatively high success rate for the Shahed drones, leading to widespread concern. However, Ukrainian defences have adapted, employing a combination of strategies to counter the threat. These include the use of air defence systems, both older Soviet-era systems and newer Western-supplied systems, such as NASAMS and IRIS-T, which have demonstrated a capacity to intercept a significant percentage of incoming Shahed drones. However, these systems are not foolproof, and the sheer volume of drone attacks continues to pose a challenge to Ukrainian air defences.

Beyond air defence systems, Ukraine has also employed improvised methods of countering the Shahed drones. Reports suggest the use of small arms fire, particularly from specially trained air defence units, has been effective in intercepting slower-moving drones. These ground-based defences highlight the adaptability and resourcefulness of the Ukrainian military in countering a novel threat with limited resources. This resourcefulness and adaptation reflect the evolutionary nature of modern conflict, where improvisation and adaptation are often as important as advanced technology.

The widespread deployment of civilian-grade drones, modified for military purposes, represents a significant aspect of the Shahed drone threat. This adaptation highlights a growing trend in drone warfare, demonstrating the ease with which readily available technology can be repurposed for military applications. The relatively low cost and accessibility of the components used in the Shahed series raise concerns about the potential for proliferation and the risk of these technologies falling into the hands of non-state actors or terrorist organizations.

Furthermore, the information warfare surrounding the Shahed drone attacks is equally significant. Russia has used the deployment of the Shahed drones to demonstrate its continued military capabilities and to project an image of resilience and determination despite battlefield setbacks. The attacks have also been used as a form of psychological warfare, aiming to undermine Ukrainian morale and to exert pressure on the Ukrainian population. In contrast, Ukraine has used footage of intercepted Shahed drones and the damage caused by them to highlight Russia's tactics and garner continued international support.

The comparative analysis of the Bayraktar TB2 and the Shahed-136 reveals contrasting approaches to drone warfare. The TB2 represents a strategy of precision and surgical strikes, while the Shahed-136 prioritizes attrition and saturation attacks. Both strategies highlight the evolving nature of modern warfare, demonstrating the adaptability of drone technology to various operational doctrines and strategic goals.

The ongoing adaptation and countermeasures employed by Ukraine against the Shahed drones provide valuable insights into the challenges of defending against large-scale, low-cost drone attacks. The Ukrainian experience underlines the need for multi-layered defence systems, combining advanced air defence technologies with adaptable tactics and innovative defensive strategies. The effectiveness of improvised methods, such as small arms fire, highlights the importance of human factors and resourcefulness in countering even sophisticated threats.

The use of the Shahed drones also raises important ethical and legal questions surrounding the use of drones in warfare, particularly regarding the targeting of civilian infrastructure. The intentional targeting of civilian infrastructure, a departure from the generally accepted principles of proportionality and distinction in warfare, has drawn widespread international condemnation. The legal implications and accountability for such attacks require further investigation and analysis. The potential for such drone attacks to destabilize critical infrastructure and generate widespread humanitarian consequences warrants attention from international organizations and policymakers.

The impact of Iranian-made Shahed drones on the Ukraine conflict extends beyond the battlefield. It underscores the evolving nature of military technology and the increasing reliance on asymmetric warfare tactics. The inexpensive and readily available nature of these drones raises concerns about the proliferation of such technology and its potential implications for future conflicts. The Shahed drone's success in inflicting damage, however indiscriminate, necessitates further analysis of defensive strategies and the development of countermeasures to mitigate the threat. This includes not just technological advancements in air defence but also the enhancement of intelligence gathering capabilities, improved coordination among different military branches, and a focus on cyber warfare, to counter the electronic components and command & control structures of these drones.

The Shahed-136's deployment has forced a reassessment of existing air defence doctrines and prompted the development of new strategies to counter the unique challenges posed by these low-cost, high-volume attacks. The experience in Ukraine highlights the need for greater international cooperation in developing effective countermeasures and addressing the broader ethical and legal implications of drone warfare. The ongoing evolution of this technological arms race underlines the critical need for continued research, development, and international dialogue to mitigate the potential risks associated with the proliferation of readily deployable and inexpensively produced drone weaponry.

The Ukrainian experience offers critical lessons in adaptability and underscores the need for a comprehensive approach to address the challenges presented by the evolving landscape of modern warfare. The case of the Shahed drones in Ukraine emphasizes that technological innovation alone is not sufficient, highlighting the critical role of adaptable military doctrine, effective intelligence, and well-coordinated responses. This narrative further complicates the discussion surrounding the future of warfare, highlighting both the risks and potential of inexpensive, mass-producible drone technology.

Beyond the headline-grabbing deployments of systems like the Bayraktar TB2 and the Shahed-136, the Ukraine conflict has witnessed a prolific and arguably more impactful proliferation of smaller, commercially available quadcopters and UAVs. These less sophisticated platforms, often modified for military use, have fundamentally altered the battlefield dynamics in ways that often go unreported in favour of the larger, more visually dramatic drone engagements. Their widespread use underscores a critical shift in modern warfare: the democratization of aerial reconnaissance and attack capabilities.

The accessibility of these smaller drones, readily available from commercial markets worldwide, has enabled both sides in the conflict – and indeed, non-state actors – to leverage aerial capabilities previously restricted to state-sponsored military programs. This accessibility has dramatically lowered the barrier to entry for deploying aerial surveillance and even offensive capabilities, drastically impacting the battlefield's information landscape and the tactical choices available to combatants. This shift has implications far beyond the immediate conflict, raising significant concerns about the potential for wider proliferation and the risks associated with placing powerful surveillance and attack capabilities in the hands of non-state actors.

One of the most prevalent uses of these smaller drones has been in reconnaissance and surveillance. Both the Ukrainian and Russian militaries, alongside various affiliated groups, use commercially available quadcopters equipped with high-resolution cameras and thermal imaging capabilities. These platforms offer real-time, localized situational awareness, proving invaluable for identifying enemy positions, tracking troop movements, assessing damage, and directing artillery fire. The low cost and ease of deployment of these drones mean they can be deployed rapidly and in large numbers, creating a distributed surveillance network that significantly enhances battlefield awareness. Their small size and relatively quiet operation provide a considerable advantage over larger, more easily detectable aircraft. The use of these smaller drones for surveillance also allows the collection of high resolution imagery of the battlefield, a form of intelligence that can be immediately integrated with other forms of reconnaissance information to provide a comprehensive picture of the situation.

Moreover, the ability to use these drones for continuous surveillance means the potential for ambush attacks is reduced, as well as the possibility of a surprise assault. The constant observation of the battlefield allows for proactive response and prevents the loss of personnel and material due to a lack of situational awareness. Furthermore, the collected data can be swiftly transmitted to command centres, accelerating the decision-making process and allowing for more efficient resource allocation on the battlefield.

However, the impact of these small drones goes far beyond simple observation. Their capacity for offensive action has also been demonstrably significant. Modified quadcopters are routinely used to deliver small munitions, improvised explosive devices (IEDs), and even grenades. These drones, while lacking the precision strike capabilities of larger drones like the Bayraktar TB2, are effective in engaging less heavily armoured targets and inflicting casualties. The relative simplicity of modifying these commercially available quadcopters, coupled with their small size and manoeuvrability, means that they are extremely difficult to detect and intercept, making them highly effective for hit-and-run attacks. The relative ease with which these drones can be modified also means that their use is likely to continue evolving, with potentially even more destructive payloads being employed as the conflict progresses.

Furthermore, the low cost of these small drones allows for a high-risk, high-volume strategy. The loss of several drones during a mission is far less impactful than the loss of a high-value aircraft such as a Bayraktar TB2. This allows for aggressive deployment strategies, overwhelming the enemy's defensive capabilities through sheer numbers. This tactical approach has been particularly effective against less mobile targets, such as artillery positions or command centres, where a swarm of drones can be used to saturate the defences and ensure the success of a strike.

The Ukrainian military has showcased particular creativity in adapting and integrating these commercially available drones into their overall strategy. Beyond their use in direct combat, these smaller drones play a central role in electronic warfare. Ukrainian forces have effectively used them to carry out electronic countermeasures (ECM) against Russian drones, creating a kind of aerial counter-drone operation. This use showcases an adaptable and innovative approach to employing readily available technology to counteract the adversary's technological advantages.

The creative use of these small drones also extends to their integration with other systems. They are frequently utilized in conjunction with artillery, providing real-time targeting information and enabling more accurate and effective strikes. This integration between human operators and unmanned systems highlights the evolving nature of modern warfare and the symbiotic relationship between human intelligence and automated capabilities.

The proliferation of these smaller, commercially available drones also raises significant questions about international law and the regulation of warfare. The ease with which they can be modified for military use, combined with their relatively low cost, makes them particularly susceptible to falling into the hands of non-state actors and terrorist groups. This raises serious concerns about the potential for these technologies to be used in conflicts around the world, outside of the controlled context of a declared war between nation-states. The lack of clear international regulations governing the use of these smaller drones in warfare further exacerbates these concerns.

The Ukraine conflict has, therefore, not only demonstrated the increasingly pivotal role of drones in modern warfare but has also showcased the surprising effectiveness and adaptability of smaller, commercially available platforms. Their use in reconnaissance, offensive attacks, and even electronic warfare underscores their growing significance in shaping battlefield dynamics. It serves as a stark reminder of the need for a comprehensive revaluation of international laws and regulations governing the use of drones, particularly in light of their growing accessibility and the consequent potential for misuse. The lessons learned in Ukraine are critical, not only for understanding the complexities of modern conflict but also for anticipating the future of drone warfare and its implications for global security.

The ease with which these technologies can be acquired and adapted demands a critical and proactive response from the international community to establish effective norms and regulations to prevent the escalation of conflict and minimize humanitarian risks. The challenges presented by this "popularization" of aerial warfare require careful consideration of international law, ethical considerations, and the development of robust countermeasures. The implications extend far beyond the immediate conflict, shaping future debates about the regulation of military technologies and the challenges posed by readily available, easily modifiable weaponry. The experience in Ukraine underscores the need for a comprehensive and adaptable approach to address the evolving dynamics of modern warfare, moving beyond simple technological competition towards a more nuanced understanding of the strategic, tactical, and ethical implications of drone warfare.

The pervasive use of drones in the Ukraine-Russia conflict has profoundly challenged established conventional warfare doctrine, forcing a re-evaluation of long-held assumptions about battlefield operations, intelligence gathering, and the overall conduct of war. The sheer scale and variety of drone deployments, ranging from sophisticated military platforms like the Bayraktar TB2 to readily available commercial quadcopters, have caused a significant adaptations in military strategy and tactics on both sides of the conflict. This impact is multifaceted, extending beyond the immediate tactical adjustments to encompass broader shifts in military planning, training, and resource allocation.

One of the most significant impacts lies in the realm of intelligence, surveillance, and reconnaissance (ISR). Traditional ISR relied heavily on manned aircraft, satellites, and ground-based intelligence networks. The introduction of drones, particularly the readily available and cost-effective smaller UAVs, has fundamentally altered this equation. The ability to deploy swarms of small drones, each capable of providing real-time video and imagery, has provided unprecedented levels of situational awareness. This constant flow of information allows commanders to make more informed decisions, react swiftly to changing battlefield conditions, and adjust their operational plans accordingly. Furthermore, the low cost of these smaller drones allows for a high-risk, high-reward approach; the loss of multiple drones is far less significant than the loss of a manned aircraft, enabling aggressive and persistent surveillance that would be cost-prohibitive with traditional methods. This shift towards a more distributed and persistent ISR capability has reduced reliance on traditional, centralized intelligence gathering methods, creating a more decentralized and agile approach to warfare.

This enhanced situational awareness has, in turn, significantly impacted tactical decision-making. The ability to identify enemy positions, movements, and equipment in real-time has facilitated more precise and effective targeting of artillery strikes, air strikes, and even ground offensives. The integration of drone imagery with other intelligence sources, such as satellite data and human intelligence, has created a more holistic understanding of the battlefield, further enhancing the effectiveness of tactical operations. The immediacy of the information provided by drones allows for quicker responses, meaning opportunities for counterattacks can be rapidly seized, whilst the ability to anticipate the enemy's manoeuvres contributes to minimizing casualties. This paradigm shift necessitates a more dynamic and adaptive approach to tactical planning, requiring military units to be capable of responding rapidly to unexpected changes on the battlefield.

Moreover, the use of drones has challenged traditional doctrines relating to air defence. The relatively small size and manoeuvrability of many drones make them difficult to detect and intercept using conventional air defence systems. This has necessitated the development of new counter-drone tactics and technologies, including the use of electronic warfare, directed energy weapons, and even other drones. The proliferation of drones has created a complex interplay between offensive and defensive capabilities, driving an arms race of sorts, where both sides continually develop and refine their strategies and technologies to maintain a counter-drone capability. This requires significant investment in technology, training, and operational doctrine to mitigate the threats posed by drones to existing air defence systems. The challenge isn't simply to shoot down drones but to disrupt and neutralize their operations. This has spurred innovation in areas such as network-centric warfare, where integrated systems can work together to track and counter drone threats.

The Ukraine conflict has also highlighted the effectiveness of drones in conducting precision strikes. While larger drones like the Bayraktar TB2 have been used for targeted strikes against high-value targets, the smaller, commercially available drones have shown their effectiveness in swarming attacks against less heavily defended targets. This capability to launch numerous attacks simultaneously, often overwhelming enemy defences through sheer volume, directly challenges traditional doctrines of attrition warfare. This tactical approach signifies a shift from aiming for large-scale, decisive battles to a more protracted conflict, where the continuous attrition of resources and manpower can have a significant cumulative effect. This new type of warfare also relies on the timely collection of information. Hence, the speed of processing and the ability to respond quickly are increasingly vital aspects of military strategy.

The integration of drones into conventional warfare has also forced a reconsideration of the role of human operators. While drones offer significant advantages in terms of range, endurance, and precision, they remain reliant on human decision-making. The need for skilled operators capable of interpreting drone imagery, coordinating strikes, and managing complex drone swarms necessitates a significant investment in training and education. This highlights a critical need for a skilled workforce capable of operating these sophisticated systems. Furthermore, the integration of AI and autonomous capabilities into drone systems raises further ethical and operational considerations, particularly around the accountability for actions taken by autonomous weapons systems.

Beyond the tactical level, the widespread use of drones in Ukraine has implications for strategic military planning. The ability to conduct reconnaissance, surveillance, and precision strikes using relatively inexpensive drones challenges the traditional reliance on large, expensive military platforms. This has implications for military budgeting, resource allocation and military industrial planning. A careful balance is needed to maintain a well-rounded and adaptable military. The strategic implications extend to the realm of asymmetric warfare, where non-state actors and even terrorist groups can leverage the availability of commercially-available drones to pose a considerable threat to larger, better-equipped militaries.

The use of drones in Ukraine has also forced a re-evaluation of the doctrine of combined arms warfare. The effective integration of drone capabilities into broader military operations necessitates a more comprehensive approach to training, communications, and command and control. Effective combined arms operations require seamless coordination among ground troops, artillery, airpower, and intelligence assets, thereby making effective communication paramount to success. Drones act as a pivotal tool in this integrated approach. The real-time intelligence and precision targeting capabilities of drones allow for more coordinated and effective strikes, optimizing the use of conventional military forces and maximizing their effectiveness on the battlefield. This integration has also highlighted the growing importance of cyber warfare and electronic warfare in the context of drone operations. The ability to disrupt enemy drone operations and protect one's own drones from attack is a key element of future military doctrine.

This conflict underscores the need for a comprehensive reassessment of international law and regulations pertaining to the use of drones in warfare. The ease with which commercially available drones can be adapted for military use and the potential for their use by non-state actors raises significant ethical and legal concerns. The development of effective international norms and regulations, coupled with national legislation, is essential to address the security implications posed by drone proliferation while upholding international humanitarian law. This issue transcends national borders and demands global cooperation to address this evolving challenge. The international community must collaborate to establish clear guidelines and regulations for the use of drones to prevent unintended escalation and reduce the risk of humanitarian crises in future conflicts.

The Ukrainian conflict has served as a brutal proving ground for drone technology, revealing both its immense potential and its inherent limitations. The lessons learned extend far beyond the immediate battlefield, impacting military doctrines, technological development, and the very nature of future conflicts. The widespread adoption of drones, ranging from sophisticated military platforms like the Bayraktar TB2 to readily available commercial quadcopters, has forced a fundamental re-evaluation of traditional warfare strategies and tactics.

One of the most significant lessons learned is the decisive role of intelligence, surveillance, and reconnaissance (ISR) in modern warfare. The Ukrainian experience demonstrated that the constant stream of real-time imagery provided by drones dramatically enhances situational awareness. This, in turn, allows for more precise targeting, improved coordination of ground and air assets, and a more responsive battlefield posture. The relatively low cost of many drone systems allows for a high-volume, high-risk approach. The loss of several inexpensive drones is far less significant financially and operationally than the loss of a manned aircraft, encouraging more aggressive ISR efforts. However, this abundance of data also presents challenges. The ability to effectively process, analyze, and disseminate this information in a timely manner is critical. Over-reliance on drone ISR without robust ground-based and human intelligence collection methods can lead to a skewed or incomplete battlefield picture. This underscores the importance of a multi-layered intelligence approach, integrating various sources to paint a complete and accurate picture of the battlefield.

The effectiveness of drones in precision strikes has also been highlighted. The Bayraktar TB2, for instance, proved capable of striking high-value targets with considerable accuracy, while swarms of smaller drones, often modified commercially available models, have proven useful in overwhelming enemy defences through sheer numbers. This capability challenges the traditional focus on large-scale, decisive battles. Instead, drone warfare seems to encourage a more protracted conflict, characterized by a continuous attrition of resources and manpower. The cumulative effect of numerous smaller, precisely targeted strikes can be devastating over time, eroding enemy morale, capabilities, and logistics. Furthermore, the adaptability of commercially available drones presents a significant challenge. Their relatively simple design makes them relatively easy to adapt for military purposes, significantly lowering the barrier to entry for non-state actors and potentially destabilizing regional security.

It has also underscored the importance of effective counter-drone strategies. The initial success of Turkish and Ukrainian forces using Bayraktar TB2 drones spurred a rapid evolution in counter-drone technologies. Russia, initially caught off guard, has since invested heavily in electronic warfare systems, directed energy weapons, and more sophisticated air defence systems specifically designed to counter UAV threats. This has triggered a technological arms race, with both sides constantly developing and refining their capabilities to neutralize enemy drones while protecting their own. The development of advanced AI algorithms to detect, identify, and track drone swarms adds a further layer of complexity to this technological competition. Successfully integrating these counter-drone technologies into existing defensive networks and adapting operational doctrines to account for these new capabilities is vital. This dynamic interplay between offensive and defensive drone technologies will likely characterize future conflicts.

However, the widespread use of drones is not without its drawbacks. The reliance on human operators, even for relatively autonomous systems, necessitates substantial investments in training and education. Highly skilled operators are required to interpret complex imagery, manage drone swarms, and make critical decisions under pressure. The potential for human error remains, and the development of more autonomous drone capabilities raises significant ethical concerns. The lack of clear legal frameworks governing the use of autonomous weapons systems requires urgent international attention. Accountability for actions taken by autonomous drones, including accidental civilian casualties, remains a highly contentious and unresolved issue.

The integration of drones into combined arms warfare has also proven critical. The seamless flow of information from drone ISR to ground forces, artillery, and air support is essential for effective coordination and maximum impact. This requires robust and secure communication networks, interoperable systems, and effective command and control structures. The Ukrainian experience has demonstrated that integrating diverse military assets with drone capabilities allows for more efficient and effective operations, minimizing friendly fire casualties and maximizing the impact of limited resources. However, the interoperability between diverse drone systems and traditional military hardware remains a significant challenge. Harmonizing communication protocols, data formats, and operational procedures across different systems is crucial for effective integration.

The strategic implications of drone warfare extend beyond the immediate battlefield. The relative affordability and ease of deployment of drones fundamentally alter the balance of power in asymmetric warfare. Non-state actors and terrorist groups can leverage readily available commercial drones to pose significant threats to much larger and better-equipped militaries. This significantly changes the calculation of military budgets and resource allocation. The increasing reliance on drones also necessitates a careful reassessment of military infrastructure, logistics, and supply chains. Ensuring the reliable and timely supply of drones, spare parts, and skilled personnel will be critical to maintaining operational effectiveness.

Furthermore, the use of drones has highlighted the increasing importance of cyber warfare and electronic warfare. The vulnerability of drone systems to hacking and electronic jamming is a significant concern. Protecting friendly drones from attack and disrupting enemy drone operations is a crucial aspect of modern warfare. This requires investment in robust cyber security measures, electronic countermeasures, and advanced network security protocols. The evolution of cyber attacks and electronic warfare specifically targeting drone systems is likely to be a prominent feature of future conflicts.

Drone Warfare in the Middle East

The shift from conventional warfare to the era of drone strikes in the Middle East marks a significant turning point in counterterrorism strategies. The United States' extensive use of unmanned aerial vehicles (UAVs), commonly known as drones, in targeting suspected terrorists and militant groups has dramatically altered the landscape of conflict, raising profound ethical, legal, and strategic questions. This approach, while lauded by some as a precise and effective tool in combating terrorism, has also been heavily criticized for its potential to cause civilian casualties and fuel anti-American sentiment. Understanding the complexities of this technology's application requires careful examination of its operational aspects, effectiveness, and long-term consequences.

The initial deployments of drones in counterterrorism operations were largely focused on gathering intelligence. The ability of UAVs to conduct persistent surveillance over vast and often inhospitable terrains provided an unprecedented advantage. Real-time imagery allowed analysts to monitor the movements and activities of suspected terrorists, gathering crucial information that could inform targeting decisions. This enhanced intelligence-gathering capacity significantly reduced the reliance on potentially unreliable human intelligence sources, increasing the precision and accuracy of subsequent strikes. However, the transition from intelligence gathering to lethal strikes was a crucial and controversial step.

The justification for U.S. drone strikes has consistently centred on the principle of self-defence, aiming to pre-empt imminent threats posed by high-value targets within terrorist organizations. The claimed precision of drone strikes minimizes collateral damage compared to traditional air strikes, although this assertion remains a subject of ongoing debate. Proponents highlight the ability to minimize civilian casualties through surgical strikes, targeting specific individuals within defined parameters, rather than unleashing widespread destruction. This precision, it is argued, limits the number of casualties, reduces the risk of civilian deaths, and minimizes the potential for powering further radicalization. Moreover, the reduced risk to U.S. personnel was touted as a significant advantage, removing the need for putting pilots in harm's way and thus diminishing the political and logistical costs of intervention.

However, this narrative is often contested. The lack of transparency surrounding the targeting process, combined with the inherent challenges in verifying the identity of targets, has led to significant criticism. The reliance on intelligence assessments, often gathered remotely, raises concerns about potential inaccuracies, biases, and the overall reliability of the information used to justify drone strikes. The reliance on algorithmic analysis of vast data sets, while aimed at enhancing precision, has also been criticized. Algorithms, though designed to be objective, may still reflect inherent biases present in the data they are trained on, potentially leading to inaccurate targeting. This lack of transparency also inhibits independent assessments of the effectiveness and ethical implications of drone operations, making it difficult to objectively evaluate the overall impact.

The debate surrounding civilian casualties remains one of the most contentious aspects of the U.S. drone program. Numerous reports and investigations by human rights organizations have documented significant numbers of civilian deaths resulting from drone strikes. The inherent difficulty in distinguishing between combatants and non-combatants in complex operational environments is often cited as a key contributing factor to civilian casualties. Furthermore, the use of lethal force in the absence of an immediate threat, even when targeting individuals who have engaged in past acts of violence, is frequently cited as a violation of international humanitarian law. The absence of due process and the lack of access to judicial review further exacerbate the ethical concerns.

Moreover, the long-term consequences of drone strikes extend far beyond immediate casualties. The perceived impunity associated with drone warfare, coupled with the often-secretive nature of the operations, has the potential to undermine the rule of law and stimulate a cycle of violence. The targeted killings, even when successful, can inspire further radicalization and recruitment within terrorist organizations. The feeling of vulnerability and the loss of trust in governments and international institutions can serve as a powerful recruiting tool for insurgent groups, perpetuating cycles of conflict. This dynamic can be especially true in environments marked by pre-existing conflict and grievances.

Furthermore, the increasing reliance on drone technology has also raised concerns about the erosion of traditional military doctrines and the potential for escalation. The relative ease and low risk of deploying drones in military operations has the potential to decrease the threshold for armed intervention, thereby potentially leading to an increase in the overall frequency of conflict. The availability of these technologically advanced weapons may alter decision-making processes, lowering the perceived risk of using force, and thus increasing the chance of unplanned military engagement.

The legal framework governing drone warfare remains uncertain and contested. The interpretation of international humanitarian law in the context of drone strikes has been a source of considerable debate among legal scholars and policymakers. The question of whether drone strikes constitute acts of war or targeted killings continues to be a subject of ongoing legal scrutiny. The lack of clear legal standards and regulations has hampered efforts to hold states accountable for the potential violations of international law resulting from drone operations. The potential for legal challenges to the use of drone technology continues to be a major consideration for states utilizing this form of military intervention.

The proliferation of drone technology has not been confined to state actors; non-state actors (NSAs) in the Middle East have rapidly integrated unmanned aerial vehicles (UAVs) into their arsenals, fundamentally altering the dynamics of conflict in the region. This shift presents unique challenges to established security paradigms and necessitates a deeper understanding of the capabilities, motivations, and strategic implications of NSA drone deployments. Initially, the use of drones by NSAs was largely limited to smaller, commercially available quadcopter systems adapted for reconnaissance and surveillance. These readily accessible devices, coupled with open-source software and readily available training materials, lowered the barrier to entry for groups lacking sophisticated technological capabilities. This enabled NSAs to conduct effective surveillance of military positions, observe troop movements, and assess vulnerabilities in security systems.

However, the evolution has been remarkable. The capacity of NSAs to acquire, modify, and deploy more sophisticated drone systems, including those capable of carrying explosives and executing targeted attacks, has rapidly increased. This development is largely attributed to the availability of readily adaptable technology from various sources, including surplus military equipment, illicit markets, and even technological expertise acquired from failed or collapsing states. The relatively low cost and ease of operation of these systems, compared to conventional weaponry, makes them an attractive option for financially constrained groups. Moreover, the decentralized nature of drone operations empowers NSAs to conduct attacks with increased efficiency and reduced risk, as the lack of centralized command structures makes them difficult to target through conventional military means.

The types of drones employed by NSAs vary considerably. While commercially available quadcopters initially formed the basis of NSA drone operations, the acquisition of more advanced military-grade systems, both through illicit means and through support from state sponsors, has significantly broadened their capabilities. This includes systems equipped with improved navigation systems, enhanced payload capacities, and improved range. The sophistication of these systems differs dramatically depending on the resources available to the NSA in question, reflecting a wide spectrum of technological capabilities. Some groups are limited to basic models used for surveillance and reconnaissance, whereas others operate drones modified to carry significant payloads of explosives.

The strategic implications of NSA drone use are profound and far-reaching. The use of drones by NSAs has introduced an unprecedented level of asymmetry in warfare. Previously, state actors held a near-monopoly on sophisticated aerial warfare capabilities. However, the democratization of drone technology has effectively levelled the playing field, allowing non-state actors to project power and conduct attacks against previously well-protected targets with relative impunity. This new capability fundamentally shifts the power balance between states and non-state actors and poses significant threats to national security, especially in the Middle East, a region already fraught with conflict and instability.

One critical aspect of NSA drone deployments is the potential for escalation. The ability of NSAs to conduct attacks against high-value targets with a relatively low level of risk can provoke retaliatory actions from state actors, potentially leading to a wider and more protracted conflict. The lack of accountability inherent in decentralized NSA organizations makes it challenging to deter such activities, further exacerbating the threat of escalation. The response by states to NSA drone attacks often involves the deployment of their own drones and more conventional military forces, potentially further destabilizing the region and contributing to a cycle of violence.

Another crucial element is the ethical and legal ramifications of NSA drone use. International humanitarian law struggles to effectively address the challenges posed by the use of drones by non-state actors. The lack of clear guidelines regarding the targeting of civilians and the verification of military objectives further complicates the issue. The use of drones by NSAs often operates outside the established framework of international law, making it extremely difficult to hold them accountable for potential violations of humanitarian norms. This challenges traditional conceptions of warfare and requires a reassessment of the existing legal framework to address these emerging threats. Accountability mechanisms are severely limited.

The challenge of countering NSA drone deployments presents significant difficulties. Traditional methods of countering aerial threats are often ineffective against smaller, more agile drones operated by NSAs. The dispersed nature of the threats makes targeting difficult, and the low cost and accessibility of drone technology make it nearly impossible to completely eliminate the threat. The need for effective countermeasures necessitates the development of multi-layered strategies involving a combination of electronic warfare systems, cyber attacks targeting drone control systems, and physical measures such as anti-drone nets and directed energy weapons. However, developing and deploying these countermeasures requires significant resources and expertise, posing a considerable challenge to states in the region.

Furthermore, intelligence gathering is essential in countering NSA drone operations. Identifying the sources of drone technology, understanding the operational methods of NSAs, and gaining insights into their organizational structures are essential to developing effective countermeasures. This includes disrupting supply chains, identifying training facilities, and neutralizing key personnel involved in the procurement, modification, and operation of drones. Cooperation among states, intelligence sharing and joint military exercises is essential in tackling this challenge.

The development of effective counter-drone technologies is also a significant area of focus. These include systems that use radio frequency jamming to disrupt drone control signals, systems that use directed energy weapons to disable or destroy drones, and systems that use artificial intelligence to identify and track drones in real-time. However, these technologies are not without limitations. Radio frequency jamming, for instance, can disrupt communications beyond the intended target, while directed energy weapons can pose safety risks. The development and deployment of these countermeasures must be done with careful consideration to the broader impacts on civilian populations and neighbouring states.

The proliferation of drone technology in the Middle East has profoundly reshaped regional power dynamics, creating new avenues for projecting power, influencing conflicts, and altering the relationships between states and non-state actors. The region, already characterized by complex geopolitical fault lines and intense rivalries, has witnessed a dramatic escalation in the use of drones, both by state and non-state actors, leading to a complex and rapidly evolving security landscape.

One significant consequence is the erosion of traditional military advantages. Historically, air power has been a decisive factor in regional conflicts, largely monopolized by major powers. The relative affordability and ease of acquisition of drone technology, however, have democratized access to aerial warfare capabilities, empowering smaller states and non-state groups to challenge established regional hegemony. This shift has led to a new era of asymmetric warfare, where weaker actors can leverage drone technology to offset the conventional military superiority of their adversaries.

The impact is particularly evident in the context of the Iran-Saudi Arabia rivalry. While both nations possess substantial conventional military capabilities, the strategic deployment of drones has added a new dimension to their ongoing proxy conflicts. Iran's support for various Shia militias in Yemen, Iraq, and Syria has been significantly enhanced by its provision of advanced drone systems, capable of conducting long-range strikes against Saudi Arabian infrastructure and military installations. These attacks, often attributed to the Houthis in Yemen or other Iranian-backed groups, have significantly hampered Saudi Arabia's military operations and exerted considerable pressure on its economy, notably targeting oil facilities. In response, Saudi Arabia has invested heavily in developing its own drone defence systems and has also undertaken retaliatory strikes. This ongoing drone-based conflict represents a crucial element in the wider geopolitical competition between these two regional powers, pushing the conflict beyond conventional military engagements into the realm of asymmetric warfare.

The involvement of external actors further complicates the situation. The United States, for example, has conducted numerous drone strikes targeting terrorist groups within the Middle East, particularly in countries like Yemen, Iraq, and Syria. While these strikes have undoubtedly inflicted casualties on terrorist organizations, they have also raised significant ethical and legal concerns regarding civilian casualties and the potential for collateral damage. The transparency and accountability surrounding these strikes remain contentious, motivated anti-American sentiment and adding another layer of complexity to the already precarious regional dynamics. Similarly, the involvement of other international actors, through the supply of arms and technology, further exacerbates the situation, creating an interconnected web of alliances and rivalries that extend far beyond the immediate region.

The role of regional powers like Turkey has also added to the complexities. Turkey's deployment of its Bayraktar TB2 drones in several conflicts, including the Libyan civil war and the Nagorno-Karabakh conflict, has demonstrated the strategic utility of these systems in influencing regional conflicts. The relative affordability and success of the Bayraktar TB2, especially in its role in assisting Azerbaijan against Armenia, has made it a highly sought-after drone system by several countries, both within and outside the region. This highlights the growing international market for drone technology and the ability of a middle-power like Turkey to leverage technological advancements for geopolitical influence.

The proliferation of drone technology in the Middle East has also heightened the risk of escalation. The ease with which drone attacks can be launched, the relative anonymity of the actors, and the often-ambiguous attribution of attacks can create a destabilizing environment, raising the potential for miscalculation and unintended consequences. The lack of clear lines of communication and the potential for rapid escalation make the situation exceptionally volatile. Each drone attack, can trigger chain reaction of retaliatory measures, potentially involving multiple actors and dragging regional partners into the conflict. This underscores the fragility of the existing peace accords and the potential for rapid escalation into larger conflicts.

The accessibility of drone technology also extends to non-state actors, and their capacity to obtain, modify, and utilize drones poses a unique threat to regional stability. The use of commercially available drones, modified for military purposes, has enabled non-state actors to conduct surveillance, target security forces, and carry out attacks against civilians. This has created a dynamic that undermines the authority of state actors, exacerbating internal conflict and creating a climate of instability.

The widespread adoption of drone warfare in the Middle East has yielded a devastating humanitarian crisis, extending far beyond the immediate battlefield casualties. The impact on civilian populations is multifaceted and profound, encompassing direct casualties, displacement, psychological trauma, and a deep erosion of trust in governing institutions. While proponents often highlight the precision targeting capabilities of drones, the reality on the ground paints a far more complex and troubling picture.

One of the most immediate and devastating consequences is the high number of civilian casualties. The inherent difficulties in accurately identifying targets in complex urban environments, coupled with the potential for errors in intelligence gathering, often result in unintended harm to non-combatants. The often-cited claim of surgical strikes often clashes with reports from human rights organizations and investigative journalists documenting significant collateral damage and the killing of innocent civilians. This discrepancy fuels distrust in official accounts and contributes to the perception of drones as indiscriminate weapons. Furthermore, the opacity surrounding drone operations, the lack of transparent investigations into civilian casualties, and the absence of meaningful accountability mechanisms exacerbate the sense of injustice and grievance among affected communities.

The lack of transparency surrounding drone strikes presents a substantial obstacle to understanding their true human cost. Governments frequently refer to national security concerns to justify the secrecy surrounding drone operations, withholding information about the targets, the methods employed, and the assessments of civilian casualties. This opacity hinders independent verification of official claims and creates a breeding ground for speculation and conspiracy theories. The absence of independent investigations into alleged civilian casualties further exacerbates the issue, leaving affected families and communities without recourse or closure. The resulting lack of accountability not only fuels resentment but also undermines the legitimacy of the operations themselves.

Beyond direct casualties, drone warfare has profound implications for the displacement of populations. The constant threat of drone strikes, often targeting specific areas or individuals, compels civilians to flee their homes in search of safety and security. This forced displacement disrupts livelihoods, shatters social structures, and adds to the immense suffering endured by already vulnerable communities. The resulting displacement contributes to the existing refugee crisis and strains already overburdened humanitarian resources. The psychological impact on those forced to abandon their homes, often with little or no warning, can be devastating, leading to long-term trauma and mental health challenges.

The psychological trauma inflicted by drone warfare transcends direct physical harm. The constant fear of unseen aerial surveillance and the ever-present threat of sudden attacks create a pervasive atmosphere of anxiety and insecurity. The psychological impact is particularly acute for those who have witnessed drone strikes firsthand or who have lost loved ones in such attacks. This trauma is further compounded by the lack of access to mental health services in many conflict-affected regions. The cumulative effect of chronic fear and the lack of support contribute to the long-term mental health challenges of entire communities, leading to social instability and hindering the prospects of recovery and reconstruction.

Furthermore, the impact of drone warfare extends to the erosion of social trust and the undermining of state authority. The perception of drone strikes as arbitrary and unjust, coupled with the lack of transparency and accountability, leads to a decline in public trust in government institutions and the rule of law. This erosion of trust can be particularly damaging in regions already grappling with political instability and conflict. It exacerbates existing social divisions, creates an environment conducive to radicalization and further violence, and hinders efforts toward reconciliation and lasting peace.

The economic consequences of drone warfare on civilian populations are equally substantial. The disruption of agricultural activities, the destruction of infrastructure, and the displacement of workers lead to widespread economic hardship and hinder long-term development. The cost of rebuilding destroyed infrastructure and providing humanitarian assistance places a significant strain on already limited resources. Furthermore, the fear of drone strikes can deter investment and hinder economic activity, leading to a vicious cycle of poverty and instability.

The use of drones in targeted killings raises complex ethical questions. Even with advanced targeting technology, the potential for misidentification and unintended consequences remains significant. The lack of due process and the absence of judicial oversight challenge fundamental principles of justice and human rights. The argument of self-defence often clashes with international humanitarian law principles, leading to heated debate on the legality and morality of drone strikes.

Addressing the humanitarian crisis necessitates a multi-pronged approach. Greater transparency and accountability in drone operations are essential to fostering trust and preventing future civilian casualties. Independent investigations into allegations of civilian harm must be conducted, and mechanisms for redress and compensation for victims and their families should be established. The strengthening of international humanitarian law to address the specific challenges posed by drone warfare is essential. The international community must work collaboratively to develop robust regulations that guarantee the protection of civilian populations while addressing legitimate security concerns.

Beyond legal and regulatory frameworks, efforts to improve intelligence gathering and targeting capabilities are critical to minimize the risk of civilian casualties. The development and implementation of improved risk assessment protocols, coupled with enhanced training for drone operators, can help mitigate the potential for errors. International cooperation in sharing information and best practices can further contribute to reducing the human cost of drone warfare.

Furthermore, significant investment in post-conflict reconstruction and recovery efforts is crucial to alleviate the long-term humanitarian consequences of drone warfare. This includes providing humanitarian assistance to displaced populations, supporting the rehabilitation of damaged infrastructure, and fostering economic development in affected regions. In addition to material support, investing in mental health services is vital to address the psychological trauma suffered by individuals and communities exposed to drone strikes.

The rapid evolution of drone technology is fundamentally reshaping the dynamics of conflict in the Middle East, promising both enhanced capabilities and new challenges. Current trends indicate a move towards greater autonomy, increased payload capacity, and improved intelligence, surveillance, and reconnaissance (ISR) capabilities. These advancements are likely to further complicate the already intricate security landscape of the region.

The Ethical and Legal Dimensions of Drone Warfare

The deployment of drones, while offering significant advantages in terms of precision and reduced risk to friendly forces, has also raised profound ethical concerns, particularly regarding civilian casualties and collateral damage. The inherent difficulties in distinguishing between combatants and non-combatants in complex urban environments or contested territories contribute significantly to this issue. The challenges are further compounded by the limitations of human intelligence gathering, the potential for flawed information, and the increasing reliance on autonomous targeting systems, raising concerns about algorithmic bias and unintended consequences.

Accurate targeting relies heavily on the quality and reliability of intelligence. Human intelligence (HUMINT), despite its importance, is susceptible to errors, misinformation, and biases. Information gathered may be outdated, incomplete, or deliberately misleading, leading to misidentification of targets and subsequent civilian casualties. The inherent difficulty in verifying the identity and actions of individuals in real-time, especially in densely populated areas, exacerbates these challenges. The reliance on visual confirmation from drone feeds, often captured from a significant distance and affected by environmental factors, can lead to mistakes, even with highly trained analysts. The inherent human limitations in interpreting complex situations under pressure contribute to this challenge.

The increasing integration of artificial intelligence (AI) in autonomous targeting systems presents another set of complex challenges. While proponents argue that AI can enhance precision by processing vast amounts of data and identifying patterns beyond human capabilities, concerns persist about potential biases within the algorithms used. These algorithms are trained on data sets which may reflect existing societal biases, potentially leading to discriminatory outcomes. For example, if a training dataset predominantly features images of males in certain contexts, an AI algorithm may be more likely to incorrectly identify a female in a similar situation as a combatant. The lack of transparency in the algorithms themselves also hinders efforts to identify and correct such biases. Moreover, the accountability mechanisms for errors made by autonomous systems are not fully developed, raising significant legal and ethical questions. Establishing clear lines of responsibility when an AI system makes an error leading to civilian casualties remains a major challenge.

The emotional and social repercussions of drone strikes on affected communities are significant and far-reaching. The sudden and unexpected nature of drone attacks, often accompanied by extreme violence, can leave deep psychological scars on survivors and witnesses. The destruction of homes and infrastructure not only disrupts daily life but can also cause significant economic hardship, exacerbating existing social inequalities. Moreover, the lack of transparency surrounding drone strikes often leads to feelings of powerlessness and resentment towards the perpetrators. This lack of accountability reinforces the sense of injustice experienced by affected populations, potentially leading to long-term instability and conflict. The loss of loved ones, compounded by the absence of traditional mourning rituals due to the sudden and often remote nature of drone strikes, further traumatizes affected families and communities.

Mitigation of civilian casualties requires a multi-faceted approach. Firstly, improved intelligence gathering is crucial. This includes enhancing HUMINT capabilities, implementing robust verification procedures, and employing advanced technologies such as high-resolution imagery analysis and predictive modelling to minimize the risk of misidentification. Stricter targeting protocols are also essential, placing greater emphasis on the verification process before authorizing strikes and establishing clear criteria for identifying legitimate targets. This could include establishing stricter chains of command and implementing thorough pre-strike assessments conducted by multiple independent analysts. These procedures should also incorporate the principle of proportionality, ensuring that the military objective justifies the risk to civilian life. Furthermore, the use of autonomous targeting systems must be carefully considered, with robust mechanisms in place to address potential biases and ensure human oversight remains central in decision-making.

Enhanced accountability mechanisms are vital to deter the reckless use of drone technology and address past wrongs. Independent investigations into alleged civilian casualties are necessary, with transparent reporting of findings and, where appropriate, appropriate consequences for those responsible. Establishing international legal frameworks and norms for drone warfare, including clearer definitions of acceptable targets and procedures for preventing civilian casualties, remains a significant challenge but is essential for addressing these issues. These measures need to include avenues for redress for victims of drone strikes, providing mechanisms for compensation and rehabilitation. International cooperation is essential in establishing common standards and procedures for drone operations to minimize the risk of civilian casualties and promote accountability.

Specific case studies highlight the complex ethical dilemmas surrounding drone strikes. The debate surrounding the use of drones in Afghanistan, Pakistan, and Yemen, as well as in the Gaza Strip, demonstrates the challenges in balancing national security interests with the protection of civilian life. In many instances, the lack of transparency surrounding these operations motivates distrust and resentment, undermining any attempts to build trust and facilitate reconciliation. The long-term consequences of these actions often extend beyond the immediate impacts of the strikes themselves, contributing to instability and resentment. A deeper analysis of these and other case studies is critical to developing effective strategies to minimize civilian casualties and promote accountability.

The development of effective countermeasures and mitigation strategies is an ongoing process. This involves also a fundamental shift in military doctrine and operational practices. A more thorough risk assessment should be conducted prior to any drone strike, weighing the potential military advantages against the foreseeable risk to civilian lives. The emphasis on precision and minimizing collateral damage must be vital in the decision-making process. Moreover, independent review boards with expertise in international humanitarian law and human rights should play a key role in evaluating the conduct of drone operations and recommending changes to minimize the risk of civilian casualties. The establishment of clear and well-defined guidelines for the operation of drones should be accompanied by the provision of comprehensive training for all personnel involved in drone operations, ensuring they understand and adhere to international law and ethical considerations.

The proliferation of drones equipped with advanced sensors and data-gathering capabilities has raised significant concerns about data privacy and surveillance. The sheer volume of data collected by military drones — from high-resolution imagery and thermal readings to geolocation data and communications intercepts — presents an unprecedented challenge to traditional notions of privacy. This data, often collected indiscriminately over vast geographical areas, can encompass not only military targets but also vast quantities of information about civilian populations, potentially including highly sensitive personal details. This raises fundamental questions about the legitimacy of such mass surveillance, the potential for its misuse, and the need for robust legal and ethical safeguards.

Existing legal frameworks governing data collection and use by military drones are often inadequate to address the unique challenges presented by this technology. International humanitarian law (IHL) provides a basic framework, but its application to drone surveillance is often ambiguous. While IHL prohibits indiscriminate attacks and requires distinction between combatants and civilians, it offers limited guidance on the permissible scope of surveillance operations. The principle of proportionality, which requires that the military advantage gained from a particular operation outweighs the harm inflicted on civilians, applies to attacks, but its implications for surveillance are less clear-cut. This ambiguity leaves significant room for interpretation and raises the possibility of violations.

National regulations vary widely, with some countries having stricter data protection laws than others. Many existing laws, however, predate the widespread use of drones and may not adequately address the unique challenges posed by this technology. For instance, laws concerning wiretapping and electronic surveillance may not fully account for the capacity of drones to collect vast amounts of data passively, without targeting specific individuals. The extraterritorial application of national laws also poses difficulties, particularly in the context of cross-border drone operations. This lack of harmonization in national and international legal frameworks creates a patchwork system, which can hinder accountability and create opportunities for abuse.

The potential for misuse of drone-collected data is substantial. This data, once collected, can be stored, analyzed, and potentially shared with third parties, raising concerns about its potential use for purposes unrelated to national security. The risk of data breaches, leading to the exposure of sensitive personal information, is also a significant concern. This sensitive data could include biometric data, facial recognition data, communications data, and location data, all of which could be used for purposes ranging from identity theft to political repression. The lack of transparency in data handling processes exacerbates these concerns, making it difficult to assess the risks and ensure accountability.

Mass surveillance through drones undermines the principle of informed consent, a base of data protection. Individuals are typically unaware that their activities are being monitored, nor do they have the opportunity to object. This lack of transparency and consent is particularly problematic when surveillance is conducted in areas with limited access to information or where dissent is suppressed. The potential chilling effect of mass surveillance on freedom of expression and association is significant. Individuals may self-censor their behaviour, fearing that their activities will be detected and tracked by drones. This can stifle legitimate political activity and restrict the ability of individuals and groups to organize and express their views.

The need for stronger oversight mechanisms to regulate drone surveillance is fundamental. This could involve establishing independent oversight bodies with the authority to review drone operations, scrutinize data collection practices, and investigate allegations of misuse. These bodies should be composed of individuals with expertise in data protection, human rights, and national security, and should be granted sufficient resources and independence to effectively perform their oversight function. Transparency measures, such as making public the legal basis for drone surveillance operations and detailing the types of data collected, are also essential to build public trust and ensure accountability.

The balance between national security interests and individual privacy rights is a delicate one. While national security considerations necessitate some level of surveillance, it is essential to ensure that these measures are proportionate and necessary to achieve legitimate security objectives. The use of drone surveillance should be subject to strict legal and ethical guidelines, carefully balancing the need for security with the protection of fundamental human rights. These guidelines should explicitly limit the scope of surveillance, specifying the types of data that can be collected, how long it can be stored, and who has access to it. Independent judicial review should be available to challenge the legality of drone surveillance operations.

Numerous instances of drone surveillance controversies highlight the urgent need for stronger regulations. In some cases, drones have been used to monitor peaceful protests and other forms of legitimate political activity, raising concerns about their use to suppress dissent. In other cases, there have been allegations of the misuse of drone-collected data, with claims that the data has been used for purposes unrelated to national security, such as targeting political opponents or conducting unwarranted surveillance of civilians. These controversies highlight the need for robust legal frameworks and effective oversight mechanisms to prevent similar incidents in the future.

Ethical considerations surrounding data collection and storage by military drones are also principal. The ethical implications of mass surveillance extend beyond the legal dimensions, raising profound questions about the fairness, transparency, and proportionality of data collection and storage practices. The potential for bias in data analysis algorithms, leading to discriminatory outcomes, is also a significant ethical concern. This bias may manifest in the targeting of specific ethnic or religious groups, potentially leading to human rights violations.

The ethical implications also extend to the issue of informed consent. As mentioned before, the lack of transparency surrounding drone surveillance means individuals are largely unaware that their activities are being monitored, thus undermining a key principle of ethical data handling. The potential for the use of this data for secondary purposes, unrelated to national security, raises additional ethical concerns. These concerns underscore the need for a comprehensive ethical framework to govern the use of drone technology, emphasizing transparency, accountability, and respect for fundamental human rights.

The Lack of International Laws and Regulation

The core principles of International Humanitarian Law (IHL) and the Laws of Armed Conflict (LOAC) – designed for traditional warfare – struggle to adequately address the unique complexities introduced by unmanned aerial vehicles (UAVs), especially those with increasing levels of autonomy. One of the most significant challenges lies in the application of fundamental IHL principles to drone operations. The principle of distinction, which mandates the differentiation between combatants and civilians, is inherently difficult to apply in the context of drone strikes. The reliance on remotely collected intelligence, often from high-altitude surveillance, presents a risk of misidentification, leading to unintended civilian casualties. The limitations of human visual identification from drone feeds, compounded by potential biases in image interpretation or the use of imperfect targeting algorithms, magnify this risk. The speed and precision of drone strikes may appear to lessen this risk, but the potential for error and the difficulty in verifying targeting information remain significant challenges. Furthermore, the potential for collateral damage, even with precise weaponry, necessitates a rigorous assessment of proportionality—that the anticipated military advantage outweighs the foreseeable harm to civilians. This assessment becomes exponentially more complex when considering the limitations of real-time situational awareness in remote operations.

Further complicating the matter is the concept of "direct participation in hostilities." Determining who constitutes a legitimate target within a complex conflict environment requires careful analysis. The lines between combatants and civilians blur in situations of asymmetric warfare, especially when non-state actors are involved. Drone operations often target individuals suspected of involvement in hostilities based on intelligence gathered through various means, creating a legal grey area where the threshold for direct participation in hostilities may not be unequivocally met. The ambiguity surrounding these legal definitions necessitates rigorous adherence to due process and the careful scrutiny of intelligence sources to ensure compliance with IHL. The lack of transparency around targeting procedures and the limitations on independent verification further exacerbate this issue.

The advent of Lethal Autonomous Weapons Systems (LAWS), also known as "killer robots," presents even more profound legal and ethical challenges. These systems possess the capacity to select and engage targets without human intervention. The prospect of machines making life-or-death decisions raises concerns about accountability, proportionality, and the very essence of human control over the use of force. The delegation of lethal decision-making to algorithms raises questions about potential biases embedded within the algorithms, and the lack of human oversight could result in unforeseen and potentially catastrophic outcomes. This presents serious hurdles to the application of IHL principles like distinction and proportionality, as the chain of command and the ability to effectively assess the circumstances surrounding a lethal engagement are significantly altered. The existing legal frameworks were not developed for scenarios where machines autonomously determine whether to take human lives, calling for urgent international discussion and the development of new regulations.

The application of LOAC is also complicated by the diffuse nature of contemporary conflicts. Drone warfare often transcends traditional battlefields, blurring the lines between war and peace. The use of drones for targeted killings outside officially declared war zones raises significant legal questions about the justification of such actions under international law. The concept of "self-defence" which underpins much of the legal justification for these strikes is open to differing interpretations and often lacks independent verification processes. Even when conducted within declared warzones, the remoteness of drone operations can lead to a detachment from the reality of human consequences, potentially undermining the moral and legal responsibility of those authorizing and executing the strikes. The increased reliance on intelligence gathering, often relying on potentially flawed human intelligence, open source data and signals intelligence, can lead to misidentification and the potential for serious violations of IHL.

The international community's efforts to address these issues remain fragmented. The UN has undertaken initiatives to codify norms and regulations surrounding the use of drones in armed conflict, but the lack of consensus among states, especially regarding LAWS, hinders significant progress. Some states champion a pre-emptive ban on LAWS, highlighting concerns about accountability and the potential for uncontrolled escalation. Others advocate for a more nuanced approach, arguing for the development of robust safety and control mechanisms before prohibiting this emerging technology. These differing viewpoints illustrate the profound complexities inherent in regulating an evolving technology that poses potentially catastrophic risks.

Legal repercussions for violations of IHL committed during drone warfare are another major concern. Investigating and prosecuting individuals for actions taken remotely presents significant logistical challenges. Determining jurisdiction, gathering evidence, and ensuring accountability can be incredibly difficult. The lack of transparency surrounding drone operations further complicates efforts to establish accountability and pursue justice. This raises concerns about the potential for impunity and the erosion of the rule of law in the context of armed conflict. International pressure and mechanisms like the International Criminal Court (ICC) hold potential for addressing such transgressions, but their effectiveness is limited by the challenges in gathering evidence and prosecuting individuals responsible for drone strikes. International cooperation and the strengthening of international legal frameworks are essential to improve accountability.

The future of drone warfare hinges on the development of comprehensive and globally accepted regulations. Addressing the legal ambiguities surrounding drone operations, including the use of LAWS, necessitates a multi-faceted approach. This includes strengthening existing IHL frameworks to explicitly address the unique characteristics of drone warfare, promoting greater transparency and accountability in drone operations, establishing independent oversight mechanisms to investigate alleged violations, and fostering international cooperation to develop common standards and regulations. The ethical considerations are deeply intertwined with the legal dimensions, requiring a comprehensive approach that prioritizes the protection of civilian lives and fundamental human rights. The ongoing technological advancements in drone technology underscore the urgent need for proactive international collaboration in shaping a regulatory framework that ensures the responsible and ethical use of this transformative technology. Failure to do so risks exacerbating the existing legal and ethical challenges, creating a pathway towards a future fraught with conflict and ethical ambiguities.

Establishing accountability in drone warfare presents unique challenges. Unlike traditional warfare, where chains of command are typically more clearly defined, the distributed nature of drone operations, involving personnel across multiple geographical locations and levels of command, complicates the process of assigning responsibility for specific actions. This is particularly true in scenarios involving autonomous or semi-autonomous systems, where the degree of human intervention in the decision-making process might be unclear. The blurring of lines between intelligence gathering, targeting decisions, and the actual execution of strikes further complicates the issue. Identifying the individuals or entities ultimately responsible for civilian casualties or other violations of international humanitarian law (IHL) becomes significantly more difficult.

One critical aspect of achieving accountability is the establishment of comprehensive and meticulous record-keeping procedures. Detailed logs documenting all stages of drone operations, including pre-strike intelligence gathering, target selection processes, the authorization chain, execution of the strike, and post-strike assessments, are essential. These logs should encompass not only technical data but also the rationale behind operational decisions, including any dissenting opinions or internal debates. This level of documentation allows for thorough post-event analysis and facilitates investigations into any alleged violations. However, implementing such systems requires a conscious and dedicated effort across all branches involved in the operation. Moreover, the secure storage and protection of such sensitive data from unauthorized access and tampering is crucial to maintain the integrity of any investigation.

Furthermore, independent investigations of alleged drone strike incidents are important for enhancing accountability. Existing mechanisms, such as internal military investigations, often lack the impartiality and transparency required to inspire public confidence. Independent investigative bodies, comprising experts from various fields, including legal, technical, and humanitarian perspectives, are necessary to conduct thorough and unbiased assessments of events. These bodies should possess the authority to access all relevant information and interview all involved parties, without fear of reprisal or political interference. The findings of these investigations should be made publicly available, fostering transparency and accountability. Such independent investigations could involve international organizations, non-governmental organizations (NGOs), or specially constituted national or international commissions, providing a layer of oversight and verification that goes beyond internal military processes. The adoption of such robust and independent oversight mechanisms would be a critical step in building trust and ensuring that actions are consistent with international law.

Improving oversight mechanisms is another critical component of establishing accountability. Current oversight frequently focuses on the ex post facto investigation of incidents, rather than proactive prevention. Enhanced oversight procedures could involve the development of clear guidelines and standard operating procedures for drone operations that explicitly address the principles of distinction, proportionality, and precaution in IHL. These guidelines could include checklists and decision matrices to assist operators in navigating complex scenarios. Regular audits and inspections of drone operations and related intelligence gathering practices can help to ensure compliance with these guidelines. Involving independent experts in these audits and inspections will further enhance the impartiality and effectiveness of the oversight process.

The lack of transparency surrounding drone warfare also undermines public trust. Governments often cite national security concerns to justify the secrecy surrounding drone operations. However, this opacity powers suspicion and speculation, making it difficult to assess the legitimacy and legality of actions undertaken. Greater transparency, while balancing legitimate security considerations, could increase public trust and confidence. This could include more detailed, though redacted, reporting of drone strikes, including the number of strikes, the locations, and general justifications. This does not imply that every detail of an operation needs to be published, but a greater level of openness could help to alleviate concerns about the secretive nature of these operations and contribute to a more informed and engaged public discourse. Furthermore, regular, transparent reporting on the number of civilian casualties and injuries, along with the methods used to minimize such harm, could help build greater trust. Establishing independent verification mechanisms, involving international organizations or neutral observers, would provide further transparency and accountability.

International cooperation is essential to addressing the accountability challenge in drone warfare. The development of globally accepted norms and regulations regarding the use of drones, especially lethal autonomous weapons systems (LAWS), is critical. Existing international frameworks, such as the UN Convention on Certain Conventional Weapons (CCW), need to be updated to explicitly address the unique characteristics of drone warfare. Such cooperation should involve the establishment of mechanisms for international investigation and monitoring of drone operations to ensure compliance with international law. The sharing of best practices and technological advancements relating to minimizing civilian harm can also improve accountability in drone operations. While national sovereignty remains vital, the need for international cooperation is evident to address the truly global nature of this form of warfare. Failure to cooperate will leave a regulatory vacuum, creating a significant risk of an unregulated arms race and a potential increase in the misuse of UAVs.

The impact of a lack of transparency on public trust cannot be overstated. When governments operate in secrecy, it creates fertile ground for speculation and mistrust. This can lead to erosion of public confidence in government institutions and the legitimacy of the use of force. It is essential to acknowledge that trust in the conduct of drone warfare is not merely a matter of public opinion; it is vital for democratic legitimacy and for preventing the escalation of conflict. Transparency and accountability are not merely desirable attributes but fundamental requirements for establishing the ethical and legal foundations for the use of UAVs in military operations. The development and implementation of robust accountability mechanisms are not only a moral imperative but a strategic necessity for maintaining international stability and security. Failure to address these issues will continue to undermine the legitimacy of drone warfare and fuel further mistrust and conflict.

International law offers a framework for addressing drone proliferation, but its application remains complex. Existing treaties, such as the Convention on Certain Conventional Weapons (CCW), offer some relevant provisions, but they were not designed to specifically address the unique challenges posed by UAVs. The CCW focuses on conventional weapons and their indiscriminate effects, and while aspects of drone technology fall under its purview, its applicability to the specific technological characteristics of drones is limited. Existing international humanitarian law (IHL), although not explicitly addressing drones, imposes restrictions on the conduct of hostilities, including the principles of distinction, proportionality, and precautions in attack. These principles are vital in regulating the use of drones to minimize civilian casualties and collateral damage. However, the translation of these principles into concrete guidelines for drone operations requires further development and clarification. The difficulty in verifying compliance with IHL in drone operations necessitates the implementation of mechanisms for monitoring and investigation.

The establishment of international norms and standards for responsible drone use is vital. This requires a collaborative effort involving states, international organizations, and experts from various fields, including technology, law, and ethics. The development of clear guidelines on the acceptable use of force, the protection of civilians, and the management of data privacy is critical. Such guidelines must consider both the military and civilian applications of drone technology, balancing the legitimate needs of national security with the concerns for human rights and international stability. The adoption of voluntary codes of conduct, while not legally binding, can play a role in promoting responsible behaviour and establishing norms within the industry. However, the effectiveness of these codes depends on widespread adoption and compliance by all relevant actors.

The legal and ethical implications of drone hacking are complex. The unauthorized access and control of a drone constitutes a criminal offense in many jurisdictions, and the consequences can range from fines to imprisonment. However, the legal framework surrounding drone hacking is still evolving, particularly with respect to the use of drones for military operations. The international law of armed conflict, as well as national laws, provides the basis for prosecuting drone hacking activities that affect national security. In addition, any incident involving injury or property damage as a direct result of hacking will likely result in far stricter criminal charges. The ethical implications are equally significant, raising questions about accountability, transparency, and the potential for misuse of technology. The development of clear legal and ethical guidelines is critical to navigating the complexities of drone hacking and ensuring that this technology is used responsibly.

The ethical considerations associated with counter-drone technologies are vital. The use of autonomous systems raises questions of accountability and the potential for unintended harm to civilians. The deployment of lethal force by autonomous systems requires careful consideration of international humanitarian law and the principles of proportionality and distinction. Robust safety mechanisms and ethical guidelines must be developed to minimize the risk of civilian casualties and ensure compliance with international norms. The establishment of clear legal frameworks is essential for regulating the use of counter-drone technologies, ensuring their responsible use, and mitigating potential risks. This includes addressing potential liability in cases of accidents or unintended consequences.

Existing international legal frameworks are ill-equipped to address the challenges posed by AWS. The Geneva Conventions and other international humanitarian law instruments were designed for a world where human actors made critical decisions in armed conflict. Adapting these frameworks to account for the complexities of autonomous weapons will require international collaboration and a willingness to address the difficult legal and ethical issues involved. The challenge lies not only in defining accountability and establishing legal responsibilities but also in articulating clear rules of engagement for autonomous systems operating in complex and dynamic environments. There is a pressing need for a robust international framework to govern the development, deployment, and use of autonomous weapons systems. This would include strict regulations on their capabilities, mechanisms for accountability, and clear standards to prevent misuse or escalation.

The path forward necessitates a comprehensive global dialogue involving experts from diverse fields – military strategists, ethicists, legal scholars, technologists, and policymakers. This dialogue must address the ethical, legal, and security implications of AWS with a view towards developing international norms and regulations to mitigate the potential risks. Open and transparent discussions are essential to create a future where the benefits of autonomous technologies are harnessed responsibly, minimizing the risks to international security and upholding humanitarian principles. Ignoring these challenges would be a grave mistake, potentially leading to an irreversible escalation in the nature and scale of armed conflict. Proactive international cooperation is the only responsible course of action, ensuring that technological advancements in weaponry do not outpace our capacity for responsible stewardship.

Existing international humanitarian law (IHL) provides a framework for regulating armed conflict. However, existing IHL principles face significant challenges in the context of autonomous weapons systems. The principles of distinction (distinguishing between combatants and civilians), proportionality (balancing military advantage against civilian harm), and precaution (taking precautions to minimize civilian harm) are difficult to implement when AI algorithms are making decisions about target selection and engagement. The inherent difficulty of verifying compliance with IHL principles when AI is involved necessitates a concerted international effort to clarify and adapt IHL to encompass the specific challenges posed by AWS. This includes the development of more precise guidelines for human oversight, verification mechanisms, and mechanisms for addressing potential violations.

International organizations play a essential role in shaping the future of drone technology and fostering international cooperation on AWS. The United Nations, through its various arms control and disarmament bodies, provides a platform for states to engage in dialogue and negotiations on arms control issues. The UN Group of Governmental Experts (GGE) on Lethal Autonomous Weapons Systems has been instrumental in facilitating discussions and exploring potential regulatory frameworks for autonomous weapons. However, the GGE's progress has been hampered by disagreements among states concerning the desirability and feasibility of a pre-emptive ban on lethal autonomous weapons.

Beyond the UN, other international organizations, such as the International Committee of the Red Cross (ICRC) and Human Rights Watch, have been actively involved in raising awareness about the ethical and legal implications of autonomous weapons. These organizations play a key role in informing the public debate, advocating for stronger regulations, and promoting adherence to IHL. Their expertise and advocacy efforts are critical for ensuring that the development and deployment of drone technology align with international humanitarian law and ethical considerations. Increased collaboration between these organizations and national governments is needed to establish a unified approach to regulating autonomous weapons systems.

Specific examples of international efforts to regulate drone technology include the Convention on Certain Conventional Weapons (CCW), which has been amended to address emerging technologies. While not explicitly focused on autonomous weapons, the CCW provides a framework for international cooperation on regulating weapon systems with potentially harmful effects. Further efforts are needed to strengthen the CCW's provisions regarding autonomous weapons and to ensure that they reflect the unique challenges posed by these technologies. This includes the development of more precise definitions of relevant terms, improved verification mechanisms, and the establishment of clearer guidelines for compliance.

Moreover, bilateral and multilateral agreements between states can play a significant role in regulating the development and deployment of drone technology. Such agreements can address specific aspects of drone technology, such as data sharing, export controls, and the establishment of technical standards. However, the effectiveness of such agreements depends on the willingness of states to comply with their provisions and on the existence of robust mechanisms for verification and enforcement. The creation of binding agreements with robust mechanisms for accountability is critical to ensuring that such bilateral or multilateral treaties are truly effective.

The development of effective arms control treaties for autonomous weapons systems requires a multifaceted approach. It necessitates international cooperation to establish clear definitions, develop robust verification mechanisms, clarify existing legal frameworks, and establish clear lines of accountability. International organizations play a decisive role in facilitating discussions, promoting compliance, and ensuring that the development and deployment of these technologies align with international humanitarian law and ethical considerations. Without such a comprehensive and collaborative approach, the risks associated with the proliferation of autonomous weapons systems will continue to grow, threatening global security and stability. The future of warfare hinges not just on technological innovation, but on the development of a robust international legal and regulatory framework capable of managing these risks effectively. The potential for both misuse and unintended consequences mandates a proactive and comprehensive global effort to establish responsible guidelines for the development and use of autonomous drone technology.

The Rise of Military Drones

The current generation of military drones reflects the culmination of these incremental advancements. Sophisticated systems like the RQ-4 Global Hawk, capable of sustained high-altitude surveillance, or the MQ-9 Reaper, with its significant payload capacity and weapons integration, represent the pinnacle of current technology. However, the evolution is far from complete. Ongoing research and development efforts continue to push the limits of drone technology, promising even more advanced capabilities in the years to come. This includes autonomous flight, swarm technologies, and advanced AI integration, all poised to revolutionize military operations and redefine the scene of modern warfare. The genesis of military UAVs, therefore, is not just a story of technological advancement, but a testament to human ingenuity and the relentless pursuit of military superiority. The journey from rudimentary target drones to sophisticated, weapon systems underscores the rapid pace of technological change and its profound impact on the nature of conflict.

The transformation of modern warfare is inextricably linked to the rise of unmanned aerial vehicles (UAVs), particularly those dedicated to Intelligence, Surveillance, and Reconnaissance (ISR). These ISR drones have revolutionized the way conflicts are planned and executed, providing a previously unattainable level of persistent surveillance and real-time battlefield awareness. Platforms such as the RQ-4 Global Hawk and the MQ-9 Reaper, when utilized in their ISR configuration, exemplify this transformative capability.

The RQ-4 Global Hawk, a high-altitude, long-endurance (HALE) UAV, stands as a testament to the advancements in unmanned aerial systems. Its ability to operate at altitudes exceeding 18,28 km provides an unparalleled vantage point, allowing for extensive area coverage and persistent observation of vast swathes of territory. Equipped with a suite of advanced sensors, including electro-optical/infrared (EO/IR) cameras, synthetic aperture radar (SAR), and signals intelligence (SIGINT) payloads, the Global Hawk can gather a comprehensive range of intelligence data, regardless of weather conditions. This persistent surveillance capability allows commanders to maintain a constant situational awareness, significantly enhancing their decision-making process. The data gathered by the Global Hawk is not merely visual; it provides a rich set of information, including detailed imagery, electronic emissions data, and even meteorological information, all contributing to a more complete understanding of the operational environment.

The MQ-9 Reaper, while often deployed in a strike capacity, plays an equally important role in ISR operations. Its extended endurance, coupled with a versatile sensor suite, allows it to provide persistent surveillance over target areas for prolonged durations. While not possessing the same high-altitude capabilities as the Global Hawk, the Reaper's lower altitude often provides greater detail and resolution in its imagery. Its ability to hang around over a target area for hours, gathering intelligence and relaying real-time data, is invaluable in supporting both offensive and defensive operations. The versatility of the Reaper's sensor suite also enables it to adapt to different operational requirements, quickly switching between EO/IR, SAR, and SIGINT modes depending on the mission parameters.

The data streams generated by these ISR drones are vast and complex, requiring sophisticated processing and analysis techniques. Ground control stations, equipped with powerful computers and specialized software, are responsible for receiving, processing, and disseminating this information. This involves not only interpreting the raw sensor data but also integrating it with other intelligence sources to create a cohesive and comprehensive picture of the battlefield. Sophisticated algorithms and artificial intelligence (AI) are increasingly being used to automate parts of this process, enhancing speed and efficiency while reducing the workload on human analysts. This automated analysis can flag potential threats, identify patterns of activity, and even predict future enemy actions, providing a significant advantage to the forces employing these systems.

However, the effectiveness of long-range surveillance is not without its limitations. Atmospheric conditions, such as cloud cover, fog, or heavy rain, can significantly impair the performance of some sensors. Similarly, electronic jamming and countermeasures can disrupt communication links and interfere with the operation of sensitive sensors. Furthermore, the sheer volume of data generated by these systems necessitates robust communication infrastructure to ensure timely and reliable transmission. This can be particularly challenging in remote or contested areas where communication networks are limited or subject to disruption. Overcoming these technological challenges remains a key area of ongoing research and development.

The ethical considerations surrounding the use of ISR drones also warrant careful consideration. The persistent surveillance capabilities of these systems raise concerns about privacy and potential abuse. The ability to monitor individuals and groups without their knowledge or consent raises serious ethical questions that need to be addressed through robust regulatory frameworks and clear operational protocols. The potential for mission creep, where ISR drones are utilized for purposes beyond their initially intended scope, also requires constant vigilance and oversight. Balancing the strategic benefits of persistent surveillance with the ethical implications is a complex challenge that requires ongoing discussion and debate among policymakers, military commanders, and ethicists.

The impact of ISR drones on operational planning and execution is profound. The availability of real-time intelligence has significantly shortened the decision-making cycle, allowing commanders to respond to evolving threats more rapidly and effectively. The ability to monitor enemy movements, identify potential vulnerabilities, and assess the effectiveness of military operations in real-time has transformed battlefield decision-making. This enhanced situational awareness reduces the risk of friendly casualties, improves targeting accuracy, and maximizes the effectiveness of military actions.

In counterterrorism operations, ISR drones have proven invaluable in identifying high-value targets, tracking their movements, and facilitating precise strikes, minimizing civilian casualties. In counterinsurgency campaigns, they have provided important intelligence on enemy movements, strength, and logistics, assisting in the planning and execution of offensive and defensive operations. During conventional warfare, ISR drones have played a critical role in providing real-time battlefield awareness, facilitating coordinated movements of ground forces, and assessing the damage caused by strikes.

The strategic impact extends beyond the immediate battlefield. The data collected by ISR drones is invaluable in informing long-term strategic planning, allowing for better resource allocation, improved force posture, and more effective policy decisions. The ability to monitor infrastructure, identify potential threats, and assess the stability of regions has transformed the way nations manage their national security interests.

The future of ISR drones promises even greater capabilities. Advancements in sensor technology, artificial intelligence, and data analytics are poised to further enhance their performance and expand their operational capabilities. The development of autonomous systems will reduce the reliance on human operators, while improved data processing techniques will facilitate more rapid and accurate analysis. The integration of multiple sensor modalities and data fusion will provide even more comprehensive situational awareness. The proliferation of smaller, more affordable drones will also enhance accessibility for a wider range of users, potentially transforming the nature of asymmetric warfare.

Drone Payload Capabilities

The effectiveness of a military drone is intrinsically linked to its payload capabilities. The payload, encompassing sensors, communication systems, and weapons, dictates the drone's operational range, its ability to gather intelligence, and its potential for offensive action. Sensors form the basis of intelligence, surveillance, and reconnaissance (ISR) missions. Electro-optical/infrared (EO/IR) sensors are ubiquitous, offering high-resolution images in both visible and infrared wavelengths. EO cameras capture visual details, while IR sensors detect heat signatures, providing valuable information even in low-light or obscured conditions. This combination is central for target identification and tracking, particularly in urban environments or at night. The quality of EO/IR imagery depends significantly on sensor resolution, zoom capabilities, and image stabilization technology. Higher-resolution sensors provide greater detail, facilitating more accurate target identification, but often come at the cost of increased payload weight and reduced flight endurance.

Synthetic aperture radar (SAR) represents another critical sensor technology. Unlike EO/IR, SAR is active, meaning it emits its own electromagnetic radiation and receives the reflections. This allows for all-weather operation, penetrating cloud cover and darkness to provide high-resolution imagery of the ground. SAR is particularly effective in identifying infrastructure, detecting camouflaged targets, and mapping terrain. Different SAR modes, such as spotlight and stripmap, offer varying levels of resolution and coverage. The selection of SAR mode depends heavily on the specific mission requirements, balancing resolution with coverage area and data processing demands. The integration of SAR and EO/IR provides a synergistic effect, offering a more comprehensive view of the operational environment. The combination provides both high-resolution visual details and all-weather ground penetration capabilities, significantly enhancing intelligence gathering capabilities.

Hyperspectral imaging represents a more advanced sensor technology, capable of capturing hundreds of narrow spectral bands across the electromagnetic spectrum. This allows for detailed spectral analysis of objects, enabling the identification of materials, vegetation types, and even chemical compounds. Hyperspectral images can be invaluable in identifying concealed targets, analyzing the composition of explosives, and monitoring environmental changes. However, hyperspectral sensors tend to be larger, heavier, and more power-hungry than EO/IR or SAR, necessitating careful consideration of the trade-offs between sensor capabilities and drone platform limitations. The processing and analysis of hyperspectral data also present significant computational challenges, requiring sophisticated algorithms and powerful processing capabilities on the ground.

Beyond these primary sensors, many drones carry additional payloads to enhance their situational awareness. These may include electronic intelligence (ELINT) sensors, which intercept and analyze radio frequency signals, or signals intelligence (SIGINT) payloads that gather communications data. These sensors provide valuable information on enemy communications, electronic warfare activities, and overall operational intentions. However, the integration of ELINT and SIGINT sensors often requires specialized antennas and signal processing equipment, adding to the overall payload weight and complexity. Furthermore, the data generated by these sensors requires specialized analysis capabilities to interpret the complex signal patterns.

The communication systems are as critical as the sensors themselves. Reliable communication is essential for data transmission, command and control, and targeting. Traditional line-of-sight communication, while simple, is limited by range and terrain. Satellite communication offers extended range but can be expensive and susceptible to jamming. Data relay networks, involving intermediate relays or manned aircraft, can significantly extend communication range and enhance reliability. The selection of communication systems is often dictated by the mission profile, operational area, and the availability of suitable communication infrastructure. In challenging environments, such as mountainous terrain or urban areas, specialized antennas and communication protocols might be needed to overcome signal attenuation and interference.

The choice of communication protocol is also critical. High-bandwidth protocols, such as advanced satellite links, enable the transmission of large amounts of data, including high-resolution images and video. However, these often require significant power and bandwidth, increasing the payload weight and power consumption. Lower bandwidth protocols may be sufficient for command and control but may compromise the transmission speed of high-resolution images.

The integration of weapons systems transforms drones from reconnaissance platforms into lethal strike capabilities. Precision-guided munitions (PGMs), such as small diameter bombs (SDBs) or laser-guided bombs (LGBs), offer the ability to engage targets with pinpoint accuracy, minimizing collateral damage. These require sophisticated targeting systems, including laser designators or GPS guidance, which must be integrated seamlessly with the drone's navigation and flight control systems. The selection of weapons depends heavily on the specific targets and the desired level of collateral damage reduction.

Air-to-ground missiles offer a different approach, with capabilities ranging from short-range anti-tank missiles to longer-range, air-to-ground weapons. These are particularly valuable for engaging moving targets or targets that are difficult to reach with PGMs. However, missiles are generally larger and heavier than PGMs, imposing significant restrictions on the drone's payload capacity and flight endurance. The choice between PGMs and missiles is often a delicate balance between precision, range, and payload limitations.

The integration of multiple payloads on a single platform presents significant engineering challenges. The trade-offs between payload capacity, flight endurance, and operational range must be carefully considered. Adding more sensors or weapons invariably increases the drone's weight, which directly impacts its flight time and operational range. Moreover, the power requirements for multiple payloads must be factored into the design, often demanding larger, more powerful batteries or a more efficient power management system.

Integrating and managing diverse payloads require sophisticated software and data fusion techniques. The data streams from multiple sensors must be seamlessly combined to provide a comprehensive and coherent picture of the operational environment. Advanced algorithms and artificial intelligence (AI) are increasingly used to automate this process, facilitating quicker analysis and faster decision-making. Effective data fusion is essential for maximizing the effectiveness of the drone's payload capabilities and extracting maximum actionable intelligence. Moreover, the effective management of multiple communication systems is essential, demanding robust protocols for data prioritization and bandwidth allocation. The integration and operation of weapon systems further demand advanced control software, reliable communication links, and fail-safe mechanisms to prevent accidents or unintended consequences.

The effectiveness of military drones hinges not only on their sophisticated sensors and weaponry but also on the ability to efficiently process and exploit the vast quantities of data they generate. Transforming raw sensor data into actionable intelligence is a complex process, requiring advanced algorithms, skilled analysts, and a robust infrastructure.

Data Acquisition and Intelligence Generation

The data overflow produced by modern military drones is amazing. A single drone equipped with multiple sensors — electro-optical/infrared (EO/IR), synthetic aperture radar (SAR), hyper spectral imagers, electronic intelligence (ELINT) and signals intelligence (SIGINT) receivers — can generate terabytes of data in a single mission. This data arrives in various formats, including high-resolution imagery, video streams, radar scans, and complex radio frequency signatures, each demanding specialized processing techniques. The initial step involves data cleaning and pre-processing, where raw data is corrected for noise, artifacts, and sensor limitations. This stage is critical for ensuring the accuracy and reliability of subsequent analysis.

Data fusion represents a critical step in transforming disparate data streams into a cohesive intelligence picture. This involves combining data from multiple sensors, often employing advanced algorithms that leverage the strengths of each sensor to overcome its limitations. For example, EO/IR imagery can provide detailed visual information, while SAR data can penetrate cloud cover and darkness. By fusing these data sources, analysts gain a comprehensive understanding of the operational environment, regardless of weather conditions or time of day. Sophisticated algorithms are employed to correlate images from different sensors, aligning geographical coordinates and identifying consistent objects or events across various data streams. This correlation helps overcome inconsistencies in individual sensor readings and ensures a more reliable assessment of the situation.

Target identification and confirmation are vital in utilizing drone-derived intelligence. Initial automated analysis may identify potential targets, but human analysts are decisive for verifying these identifications, taking into account context, patterns, and other intelligence sources. This process involves meticulous scrutiny of the images and data, cross-referencing with other intelligence reports, and considering operational context. In situations where armed intervention is being considered, target confirmation is decisive for preventing friendly fire incidents and minimizing collateral damage. This process emphasizes the importance of integrating drone data into a broader intelligence framework, corroborating information from other sources to ensure a high degree of confidence in target identification.

The integration of drone data with other intelligence sources significantly enhances the overall situational awareness. Drone-derived images and sensor data are frequently integrated with information from human intelligence (HUMINT), signals intelligence (SIGINT), and other sources to create a holistic understanding of the battlefield. This data fusion approach helps corroborate information, identify patterns, and form a comprehensive operational picture. Databases and analytical platforms are essential for managing this integration, enabling analysts to efficiently query, filter, and analyze data from multiple sources.

The role of artificial intelligence (AI) and machine learning (ML) in data analysis and exploitation is rapidly growing. AI algorithms are increasingly used to automate many aspects of data processing, analysis, and target recognition. This automation reduces human workload, accelerates analysis, and potentially improves the accuracy of intelligence assessments. AI algorithms can be trained to identify patterns, anomalies, and objects of interest in vast quantities of images and data that would be overwhelming for human analysts. For example, AI can be trained to identify specific types of vehicles, weapons systems, or infrastructure, facilitating quicker identification and assessment of threats. However, human oversight remains critical, especially in sensitive operational contexts. The human-machine partnership, balancing AI capabilities with human judgment, provides the most effective approach to data exploitation.

The impact of drone-derived data analysis on real-world military operations is evidently significant. In numerous conflicts, drone data has been instrumental in guiding precision strikes, supporting troops on the ground, and gathering decisive intelligence on enemy movements and capabilities. The ability to conduct persistent surveillance over large areas, using drones as persistent ISR platforms, has provided invaluable insights into enemy activity, assisting in the planning and execution of operations. The real-time nature of drone data allows for immediate response to developing situations, enhancing situational awareness and response time for military commanders. Furthermore, the ability to analyze large datasets retrospectively has proven invaluable for understanding the evolution of conflict, identifying trends, and extracting long-term intelligence insights.

However, challenges remain in the efficient and effective exploitation of drone data. The sheer volume of data can overwhelm existing processing capabilities, requiring significant investment in infrastructure and processing power. Data security and protection are also critical concerns, requiring robust systems to protect sensitive information from unauthorized access or manipulation. The development and maintenance of effective data analysis procedures and skilled analysts are also essential for maximizing the value of this data. The development of standardized data formats and interfaces is vital for seamless integration between different drone platforms and intelligence systems.

The future of data analysis and exploitation in drone operations will be characterized by further advancements in AI and ML. As these technologies continue to evolve, the automation of data processing and analysis will increase, leading to faster and more efficient intelligence production. Real-time processing and analytics capabilities will become even more critical, enabling more rapid responses to developing situations. The integration of advanced sensor technologies, such as quantum sensors, will further enhance the quality and quantity of data generated by drones, leading to a new level of situational awareness and operational effectiveness.

The proliferation of military drones has ushered in a new era of warfare, characterized by unprecedented capabilities in surveillance, precision strikes, and asymmetric conflict. However, this technological revolution has also raised profound ethical and legal questions that demand careful consideration. The very nature of drone warfare — the ability to conduct strikes from thousands of miles away, often with limited human oversight — challenges traditional notions of accountability, transparency, and the jus ad bellum (right to go to war) and jus in bello (laws of war) principles enshrined in international law.

One of the most pressing concerns is the issue of accountability. When a drone strike results in civilian casualties, determining responsibility can be exceptionally difficult. The physical distance between the operator and the target, often coupled with the use of sophisticated targeting algorithms and AI-assisted decision-making, creates a degree of operational detachment that blurs lines of responsibility. Is the pilot responsible? The intelligence analyst who identified the target? The policymakers who authorized the strike? The programmers who designed the targeting software? Establishing clear lines of accountability in such complex scenarios is a major challenge, demanding a more robust legal framework than currently exists. Current international humanitarian law struggles to adequately address these complexities, often relying on vague concepts like "proportionality" and "distinction" that are challenging to apply consistently in the fast-paced world of drone operations.

The lack of transparency surrounding drone operations further exacerbates the accountability problem. Many countries operating drone programs maintain strict secrecy about their targets, the number of strikes conducted, and the associated civilian casualties. This opacity hinders independent verification of compliance with international humanitarian law, allowing for potential abuses to go unchecked. Transparency is essential not only for holding states accountable but also for fostering public trust and confidence in the legitimacy of drone operations. Independent investigations and the establishment of international mechanisms for monitoring drone programs are crucial steps toward achieving greater transparency.

The potential for civilian casualties is another major ethical and legal concern. While proponents argue that drones offer greater precision than traditional air strikes, the reality is that drone strikes have resulted in significant civilian casualties. These incidents often stem from a confluence of factors, including faulty intelligence, limitations in sensor technology, and the challenges of identifying combatants in complex urban environments. The issue extends beyond simply counting civilian casualties. The increasing reliance on autonomous weapons systems (AWS), also known as "killer robots," presents an even greater challenge to accountability and transparency. The delegation of life-or-death decisions to machines raises fundamental questions about human control, responsibility, and the very nature of warfare. The potential for unforeseen consequences, algorithmic biases, and the erosion of human judgment in decision-making processes demand a cautious consideration of the ethical and legal implications of autonomous weapons before widespread deployment. Several nations have already called for a pre-emptive ban on such technologies, recognizing the potential for catastrophic consequences.

Combat and Armed Drones

Unmanned Combat Aerial Vehicles (UCAVs) represent a significant change in the scene of modern warfare. Their capabilities, while impressive, are not without limitations, and their operational doctrines are constantly evolving to address both technological advancements and the realities of contemporary conflict. The deployment of UCAVs has fundamentally altered the dynamics of precision strikes, close air support, and numerous other combat roles.

One of the most prominent examples of a UCAV is the MQ-9 Reaper. Its endurance, payload capacity, and sophisticated sensor suite make it a versatile asset for a wide range of missions. Equipped with Hellfire missiles and other guided munitions, the Reaper has demonstrated its effectiveness in conducting precision strikes against high-value targets, often in challenging environments where manned aircraft would face significantly higher risks. Its ability to hang around over target areas for extended periods, providing persistent surveillance and enabling real-time assessment of the situation before engagement, significantly enhances operational effectiveness. The Reaper's operational history, including extensive use in Iraq, Afghanistan, and other conflict zones, provides a wealth of data for analyzing its capabilities and limitations in real-world scenarios. Data from these deployments reveals not only the successes, but also the challenges faced in maintaining situational awareness, dealing with electronic countermeasures, and minimizing collateral damage.

The Bayraktar TB2, a smaller and more affordable UCAV, offers a compelling alternative, particularly for countries with less robust air power capabilities. Its relatively low cost and ease of operation have contributed to its widespread adoption by several nations. While its payload capacity is lower than the Reaper's, its effectiveness in providing close air support and conducting precision strikes has been demonstrated in several conflicts, most notably in the Nagorno-Karabakh war. However, the TB2's smaller size also means that it has a shorter range and is more vulnerable to air defence systems. Analysis of its operational performance highlights the importance of considering the specific operational environment and the nature of the threats faced when deploying a UCAV. The TB2's success also underscores the trend of asymmetric warfare, where less technologically advanced nations can leverage readily available, yet effective, technology to level the playing field.

The Iranian Shahed-136, often categorized as a loitering munitions or "kamikaze drone," represents a different type of UCAV. Unlike the Reaper or TB2, which are designed for repeated use, the Shahed-136 is a one-way weapon, programmed to impact its target. This low-cost, expendable design creates a considerable advantage in sheer volume of fire, potentially saturating enemy defences. The Shahed-136's operational deployments in Ukraine, particularly the sheer number employed, demonstrate the potential impact of employing large swarms of inexpensive UCAVs. However, their limited precision and susceptibility to countermeasures limit their overall tactical effectiveness against sophisticated defences. The Shahed-136's impact, while concerning, highlights the need for robust air defence systems, especially those designed to counter drone swarms.

Analyzing the operational doctrine employed with UCAVs reveals several key aspects. The reliance on intelligence, surveillance, and reconnaissance (ISR) capabilities to identify and locate targets is vital. This often involves a networked approach, integrating data from various sources, including satellites, manned aircraft, and ground-based sensors, to create a comprehensive picture of the battlefield. The selection of appropriate munitions is also vital, depending on the target and the desired level of precision. The integration of sophisticated targeting systems and AI-assisted decision-making tools has significantly enhanced the accuracy and effectiveness of UCAV strikes, reducing the risk of collateral damage. However, there are clear limitations to the reliance on technology, as imperfect data or algorithm biases can lead to significant errors.

The operational limitations of UCAVs are equally important to consider. Their dependence on satellite communication links, for instance, makes them vulnerable to jamming or disruption, potentially hindering their effectiveness. Furthermore, their vulnerability to air defence systems, particularly sophisticated surface-to-air missile systems, remains a significant concern. The increasing development of advanced air defence systems specifically targeting drones presents a formidable challenge to UCAV operations. This necessitates the development of tactics and technologies to mitigate these risks, including the use of electronic countermeasures, advanced manoeuvring techniques, and the deployment of UCAVs in coordinated swarms to overwhelm enemy defences.

Furthermore, the operational environment significantly impacts the effectiveness of UCAVs. Urban environments, for example, pose particular challenges due to the difficulty of distinguishing between combatants and civilians, increasing the risk of collateral damage. Adverse weather conditions, including fog, rain, and snow, can also severely limit the effectiveness of UCAV sensors and navigation systems. The development of all-weather UCAVs is critical to expand the operational envelope and enhance their overall utility across various environments.

Another key aspect of UCAV operations is the integration with manned aircraft and ground forces. This coordinated approach, commonly referred to as "multi-domain operations," can significantly enhance the effectiveness of UCAVs by providing real-time situational awareness, target acquisition support, and close coordination between different elements of a military force. This synergy between UCAVs and other assets is critical to maximizing their potential in diverse combat scenarios. The effectiveness of such integrated approaches depends on robust communication networks and the ability to share information seamlessly between different platforms and units.

The future of UCAVs in modern warfare will likely involve further advancements in their capabilities, including increased autonomy, improved sensor technology, enhanced resilience to air defence systems, and greater integration with other military platforms. The development of swarm tactics, where large numbers of relatively inexpensive UCAVs are employed in coordinated attacks, presents both opportunities and challenges. While this approach can potentially overwhelm enemy defences, it also necessitates the development of new countermeasures and operational doctrines. The integration of AI and machine learning will play a pivotal role in enhancing their autonomous capabilities, but also raises critical questions regarding accountability and the potential for unintended consequences.

Precision strikes, a basis of modern military strategy, have undergone a dramatic transformation with the advent of armed drones. These unmanned aerial vehicles (UAVs), or Unmanned Combat Aerial Vehicles (UCAVs), offer a unique blend of capabilities, enabling pinpoint accuracy while minimizing – at least theoretically – the risk to friendly forces. However, the reality of achieving truly collateral-damage-free precision strikes is complex and multifaceted, involving a sophisticated interplay of technology, intelligence, and operational doctrine.

Targeting techniques employed by armed drones are remarkably diverse and depend heavily on the specific mission, the target's characteristics, and the available intelligence. Satellite images, aerial reconnaissance from manned or unmanned platforms, and human intelligence (HUMINT) all contribute to the creation of a detailed target profile. This process, often involving a lengthy chain of command and meticulous scrutiny, aims to ensure that the target is accurately identified and its location precisely pinpointed. Sophisticated algorithms and image recognition software play an important role in analyzing this information, helping to distinguish between military and civilian infrastructure, and even individual combatants and non-combatants within complex environments. This process, while highly advanced, is not foolproof and remains vulnerable to human error or misleading intelligence.

The weapon systems used in precision strikes by armed drones vary depending on the target and the mission objectives. Hellfire missiles, a mainstay of many UCAV arsenals, are guided munitions capable of striking moving targets with considerable accuracy. Their laser-guided and millimetre-wave radar capabilities allow for effective targeting in various weather conditions, though atmospheric interference can still impact precision. Other guided munitions, including small diameter bombs (SDBs) and precision-guided bombs (PGBs), offer different capabilities. SDBs, due to their smaller size, can be used in more confined spaces, while PGBs offer greater destructive power for larger targets, although they are generally less agile. The selection of the appropriate weapon system requires careful consideration of the target's size, location, and surrounding environment, aiming to minimize collateral damage.

However, the effectiveness of these weapons systems is intrinsically linked to the accuracy of the targeting data. Even with advanced guidance systems, factors such as atmospheric conditions, electronic jamming, and the target's movement can influence the accuracy of the strike. Adverse weather conditions, such as fog, rain, or snow, can significantly degrade sensor performance, reducing the accuracy of targeting and increasing the risk of collateral damage. Similarly, electronic countermeasures employed by the enemy can disrupt the guidance systems of the munitions, leading to inaccurate strikes or complete mission failure. The inherent limitations of the sensor technology themselves, such as resolution limitations in satellite images or the challenges of distinguishing between combatants and non-combatants in crowded urban areas, contribute to the difficulty of achieving flawless precision.

The minimization of collateral damage is a main concern in all precision strike operations. The potential for civilian casualties, even with the use of advanced targeting technologies, remains a significant ethical and legal challenge. International humanitarian law emphasizes the distinction between combatants and civilians, and the principle of proportionality, which dictates that the anticipated military advantage must outweigh the expected harm to civilians. However, the application of these principles in real-world scenarios is often complex and challenging.

The increasing autonomy of armed drones further complicates the issue of collateral damage. While autonomous systems can potentially enhance the speed and efficiency of targeting decisions, they also raise concerns about accountability and the potential for algorithmic bias. The development and deployment of autonomous weapons systems (AWS) necessitates careful consideration of ethical and legal implications, demanding stringent regulations and international cooperation to ensure responsible use and prevent unintended escalation. The "human-in-the-loop" approach, while aiming for greater control, faces technological limitations in communication delays and the inherent stresses of real-time decision-making. Striking a balance between autonomy and human oversight remains a significant challenge.

The evolving nature of warfare, with the proliferation of sophisticated air defence systems and the emergence of new tactics and technologies, necessitates a continuous adaptation in precision strike capabilities. The development of countermeasures to defeat air defences, improved sensor technology to operate effectively in challenging environments, and more robust communication systems to minimize latency are all critical areas of ongoing research and development. The successful deployment of precision strikes involves the seamless integration of intelligence, surveillance, reconnaissance (ISR) assets, sophisticated targeting systems, and appropriately selected weapon systems, backed by clear rules of engagement and robust post-strike analysis. This intricate process demands constant evaluation and improvement to adapt to the ever-changing dynamics of modern conflict.

Close air support (CAS) operations represent a critical and demanding application of armed drone technology. The integration of unmanned aerial vehicles (UAVs) into this domain, however, presents a unique set of challenges that go beyond the complexities of precision strikes against stationary targets. The dynamic, fluid nature of the battlefield, the need for rapid response times, and the essential imperative to avoid friendly fire necessitate a profound shift in operational doctrine and technological capabilities.

Effective CAS operations with drones hinge on seamless communication and coordination between airborne assets and ground troops. Real-time information exchange is vital, enabling the drone operator to gain an accurate picture of the battlefield situation, identify enemy positions, and differentiate between friend and foe. This necessitates robust and secure communication links, often relying on a combination of satellite and terrestrial networks, to overcome the challenges of battlefield interference and enemy jamming efforts. The latency of communication, even with advanced systems, remains a critical factor influencing the responsiveness of drone-based CAS, necessitating efficient data processing and quick decision-making algorithms.

Target identification in the dynamic context of CAS is exceptionally demanding. Ground forces may be operating in close proximity to civilian populations, creating a high risk of collateral damage. The drone operator must possess exceptional situational awareness, relying on a multitude of sensors, including high-resolution electro-optical/infrared (EO/IR) cameras, synthetic aperture radar (SAR), and potentially even signals intelligence (SIGINT) to accurately assess the situation and pinpoint the hostile elements while mitigating the risk to non-combatants. The integration of human intelligence (HUMINT) from ground forces provides invaluable context for the drone operator, allowing them to cross-reference information and validate target identification before initiating an attack.

The use of digital battlefield management systems, which integrate data from multiple sources, including manned and unmanned platforms, and provide a common operational picture, are central for effective CAS. These systems provide ground commanders and drone operators with real-time situational awareness, aiding in target acquisition, minimizing the risk of friendly fire, and improving coordination between different units. However, the interoperability of these systems across different platforms and services remains a significant challenge, requiring standardization efforts and interoperability agreements to ensure seamless data exchange.

Avoiding friendly fire is of supreme importance in CAS operations. The close proximity of ground troops to potential targets requires exceptional precision and meticulous planning. Strict rules of engagement (ROE), clearly defining the conditions for engagement and specifying the types of targets permitted to be struck, must be meticulously followed. Furthermore, the use of advanced targeting systems, such as laser designation from ground troops, enhances the accuracy of the strike, minimizing the risk of collateral damage and accidental strikes on friendly forces. The implementation of robust "positive identification" protocols, involving multiple confirmation steps before engagement, is also essential to mitigate the risk of misidentification.

The effectiveness of drone-based CAS varies depending on the specific characteristics of the operational environment and the nature of the targets. In open terrain, where clear lines of sight are available and the risk of civilian presence is minimal, drones provide an effective and highly precise method of delivering CAS. However, in densely populated urban areas or complex terrains, the risk of collateral damage is significantly higher, requiring more cautious and deliberative targeting procedures. The use of smaller, less powerful munitions may be preferred in these contexts to mitigate the impact of unintended strikes. Furthermore, the presence of effective enemy air defences, including man-portable air defence systems (MANPADS), poses a considerable threat to drones conducting CAS, requiring the use of advanced countermeasures and careful route planning.

The development of autonomous targeting systems promises to further enhance the capabilities of drone-based CAS. However, this technological advancement raises considerable ethical and legal implications, particularly concerning the potential for algorithmic bias and the challenge of maintaining human oversight. The use of AI in CAS necessitates a careful balancing act between maximizing operational efficiency and minimizing the risk of unintended harm. The "human-in-the-loop" model, wherein humans retain ultimate control over the engagement decision, remains a critical requirement, though the development of more robust human-machine interfaces and faster communication protocols is essential to improve the responsiveness of the system.

The asymmetric nature of modern warfare, characterized by conflicts between actors with vastly different military capabilities, presents a unique operational landscape for unmanned combat aerial vehicles (UCAVs). These conflicts, often involving non-state actors, irregular forces, and technologically less-advanced militaries, create both opportunities and challenges for the deployment of drones. The advantages offered by UCAVs in such scenarios are compelling, but understanding their limitations is equally decisive for effective operational planning and strategic decision-making.

One of the most significant advantages of UCAVs in asymmetric warfare lies in their cost-effectiveness. Compared to manned aircraft, drones offer a significantly lower operational cost per sortie. This is particularly relevant in prolonged conflicts or situations where resources are scarce. The reduced risk to human pilots also contributes to lower overall costs, as expensive pilot training, life insurance, and potential rescue operations are largely eliminated. This cost-effectiveness allows for a higher operational tempo, enabling sustained surveillance, reconnaissance, and strike capabilities that may be beyond the reach of conventional air forces with limited budgets. The reduced risk to personnel is another key advantage. The inherent danger of manned flight in hostile environments, particularly those characterized by intense ground fire, sophisticated air defences, or the presence of advanced anti-aircraft weaponry, is significantly mitigated with UCAVs. This allows for operations in high-risk zones where deploying manned aircraft would be impractical or excessively dangerous. This is especially important in asymmetric warfare, where adversaries may rely on guerrilla tactics, improvised explosive devices (IEDs), or sophisticated air defence systems that are disproportionate to their overall military capabilities. The ability to conduct persistent surveillance and reconnaissance without jeopardizing human life is a considerable advantage in such scenarios, enabling a more comprehensive understanding of the battlefield and minimizing friendly casualties.

The precision strike capabilities of UCAVs are particularly valuable in asymmetric warfare. The ability to deliver pinpoint strikes against specific targets, minimizing collateral damage, is essential in environments where civilian populations might be inadvertently affected. This precision is especially vital in densely populated urban areas or when engaging targets located near critical infrastructure. While the potential for civilian casualties remains a central ethical concern, technological advancements in targeting systems and the incorporation of sophisticated algorithms for risk assessment are constantly striving to improve the accuracy and minimize unintended harm. The utilization of advanced sensors like electro-optical/infrared (EO/IR) cameras, synthetic aperture radar (SAR), and multispectral imaging enhances the ability to discriminate between combatants and non-combatants, improving the precision of the strike.

However, the deployment of UCAVs in asymmetric warfare is not without its disadvantages. One critical limitation is their vulnerability to sophisticated air defences. Although UCAVs possess several advantages over manned aircraft, such as reduced radar signatures and the ability to manoeuvre unpredictably, they are still vulnerable to more advanced air defence systems, particularly man-portable air-defence systems (MANPADS). The increasing proliferation of effective MANPADS in several regions presents a significant threat, necessitating careful operational planning and the implementation of countermeasures like electronic warfare systems and sophisticated jamming techniques. The loss of a UCAV, while less costly than the loss of a manned aircraft and its pilot, can still represent a significant loss of capabilities and intelligence.

Another significant challenge relates to the potential for escalation. The relatively low cost and ease of deployment of UCAVs can inadvertently lower the threshold for military intervention. The reduced risk to personnel might encourage more frequent and less carefully considered uses of force, potentially escalating conflicts that could have otherwise been resolved through other means. The perception of drones as being somehow less lethal than manned aircraft can be misleading, contributing to the risk of unintentional escalation. This necessitates a rigorous framework for their deployment, incorporating strict rules of engagement, clear lines of authority, and robust political oversight.

Moreover, the reliance on technological superiority can be a double-edged sword. Asymmetric warfare is often characterized by the adversary's ability to adapt and exploit vulnerabilities. This requires constant adaptation and innovation to anticipate and mitigate emerging threats. A successful tactic may become obsolete rapidly as adversaries develop countermeasures or adopt novel techniques. The potential for cyber attacks and electronic warfare against UCAV control systems also poses a significant threat, underscoring the importance of robust cyber security measures and the ability to maintain effective communication links under duress.

The reliance on satellite-based communication links can also create vulnerabilities. The effectiveness of UCAVs is heavily dependent on satellite communication for data transmission, control, and targeting information. Disruptions to these links, through jamming or other means, can significantly impair operational capabilities, leading to loss of control or inability to complete missions. Therefore, contingency plans should encompass alternative communication methods, emphasizing resilience and redundancy in communication networks.

The Future of Drones

The future of drone warfare is likely to be characterized by increased sophistication, autonomy, and integration with other military technologies. The development of swarm technology, which allows for the coordinated operation of multiple drones, promises to transform the nature of warfare. The integration of drones into larger military systems will further enhance their effectiveness and lethality. The continuous advancements in drone technology necessitate a continuous reassessment of operational strategies and ethical implications, highlighting the need for international cooperation and dialogue to address these significant challenges.

One of the most significant developments will be the continued miniaturization of drone technology. The emergence of smaller, more agile UAVs capable of operating in diverse and challenging environments is a reality. This trend is likely to accelerate, resulting in drones that are virtually indistinguishable from birds or insects. Such miniature drones, equipped with advanced sensors and potentially even miniature weapons systems, could be deployed for covert surveillance, reconnaissance, and even targeted assassinations with a dramatically reduced risk of detection. This capability raises serious concerns about privacy and the potential for escalation of conflicts, particularly in urban environments where distinguishing combatants from civilians becomes increasingly difficult. The ability to deploy swarms of these micro-drones, each carrying a small payload, could also overwhelm traditional air defence systems and create a new level of asymmetric warfare capability.

Another crucial factor in the future of drone warfare will be the increasing levels of autonomy. While current drones often require human operators for decision-making, particularly regarding targeting, there is a clear push towards greater autonomy, driven by the desire for faster response times and reduced reliance on human operators in dangerous or remote locations. This trend toward autonomous or semi-autonomous systems poses significant ethical and legal challenges. The delegation of life-or-death decisions to algorithms raises serious questions about accountability, particularly in cases of unintended casualties or malfunctions. Determining responsibility for actions taken by autonomous drones — about who is legally liable when a malfunction leads to civilian casualties — will be a significant legal difficulty requiring international cooperation and new legal frameworks. Furthermore, the risk of unintended escalation is amplified by the reduced human involvement in decision-making, introducing a potential for algorithmic biases or unforeseen consequences not adequately accounted for by human oversight.

The development of swarm technology represents a further significant evolution in drone warfare. Swarms of coordinated drones, communicating and cooperating in real time, could overwhelm enemy defences, conduct complex manoeuvres, and accomplish missions beyond the capabilities of individual drones. A swarm of small, relatively inexpensive drones overwhelming air defences, saturating enemy radar systems, or conducting a coordinated attack on a specific target. This capability presents a significant technological shift, potentially disrupting existing military doctrines and necessitating a reassessment of defence strategies. The potential for using AI-driven algorithms to manage these swarms further complicates matters, raising questions about the predictability and controllability of such systems, particularly in chaotic combat environments. Moreover, the potential for misuse of swarm technology, both by state and non-state actors, is a serious consideration. This technology could empower smaller, less technologically advanced actors to engage in large-scale attacks, significantly destabilizing existing power dynamics.

The integration of artificial intelligence (AI) into drone technology will dramatically reshape drone warfare. AI-powered drones can enhance situational awareness, improve targeting accuracy, and automate various aspects of drone operation, from navigation to target acquisition and engagement. However, the use of AI in lethal autonomous weapons systems (LAWS) raises profound ethical and legal questions. These systems raise concerns about the potential for unintended consequences, algorithmic bias, and the erosion of human control over lethal force. The potential for autonomous drones to make life-or-death decisions without direct human intervention raises fundamental questions about accountability and the very nature of warfare. The difficulty in predicting and controlling the behaviour of sophisticated AI algorithms in unpredictable combat environments poses a serious risk of unintended escalation. Furthermore, the potential for AI-driven drones to learn and adapt, potentially in unexpected ways, further complicates the assessment of risks and the development of appropriate regulations.

These technological advancements also have profound implications for international relations. The proliferation of drone technology, particularly to non-state actors, could exacerbate existing conflicts and destabilize regions. The relative affordability and ease of use of some drone systems could empower non-state actors to challenge state power in ways previously unimaginable. The potential for the use of drones in cyber warfare further complicates the picture, introducing the risk of sophisticated attacks on critical infrastructure or other sensitive systems. The absence of a comprehensive international regulatory framework governing the development, deployment, and use of drones exacerbates these risks, increasing the likelihood of an uncontrolled arms race and the potential misuse of these technologies.

The future of drone warfare, therefore, presents a complex and multifaceted challenge. The predictions outlined above highlight the need for proactive, international cooperation to establish clear ethical and legal frameworks that govern the development, deployment, and use of this rapidly evolving technology. International agreements, analogous to the existing international treaties regulating conventional weapons, are urgently needed to address the unique challenges posed by autonomous drones, swarm technology, and AI-powered systems. Failure to address these issues proactively will likely result in an unregulated arms race, potentially leading to increased instability, escalating conflicts, and an unacceptable level of civilian casualties. The development of robust oversight mechanisms, enhanced transparency, and impartial investigative processes are all crucial elements in mitigating the risks associated with the future of drone warfare. Ignoring these considerations risks not only ethical violations but also a significant deterioration of global security. The path forward demands a concerted effort from governments, international organizations, and technology developers to ensure that this powerful technology is used responsibly and ethically, promoting security and stability rather than powering conflict and instability. This requires a global dialogue not merely about technological advancement but about the very nature of warfare in the 21st century and how to manage the inherent risks of increasingly autonomous weapons systems. The stakes are high, and the time for decisive action is now.

Drones Technology

The global spread of drone technology among nation-states is driven by a confluence of factors. Cost-effectiveness stands out as a primary motivator. Compared to traditional manned aircraft, drones offer a significantly lower acquisition and operational cost. This is especially attractive to nations with limited defence budgets, allowing them to access advanced surveillance and strike capabilities previously beyond their reach. This cost advantage is particularly pronounced when considering the cost of training and maintaining skilled pilots, a substantial expense avoided with drone deployments. The increasing affordability of commercially available drone components further stimulates this trend, lowering the barrier to entry for states seeking to develop or acquire their own UAV capabilities.

Technological advancements are another key driver. The rapid pace of innovation in drone technology, encompassing improved sensors, enhanced autonomy, and more sophisticated weapon systems, creates a constant demand for upgrades and new acquisitions. States strive to maintain technological parity or even achieve a competitive edge in this rapidly evolving domain. This relentless pursuit of technological superiority leads to a continuous cycle of acquisition, upgrades, and the development of countermeasures, effectively motivating an arms race in the realm of unmanned aerial vehicles.

Geopolitical considerations also heavily influence drone acquisition patterns. Regional conflicts, proxy wars, and broader power struggles all contribute to the demand for drones. States facing internal or external threats, asymmetric warfare scenarios, or territorial disputes find drones to be valuable assets for surveillance, reconnaissance, and targeted strikes. The potential for rapid deployment, minimizing risk to personnel, makes drones particularly appealing for operations in hostile or contested territories. Furthermore, the ability of drones to operate discreetly can be a significant advantage in managing international tensions and avoiding direct military confrontation.

The types of drones acquired vary significantly depending on the acquiring nation's specific needs and capabilities. Smaller, less expensive drones are frequently used for surveillance and reconnaissance missions, offering valuable intelligence-gathering capabilities without the high risks associated with manned aircraft. These smaller systems are relatively easily deployed and maintained, making them adaptable to a wide range of operational scenarios. In contrast, larger, more sophisticated drones, often equipped with precision-guided munitions, are typically employed for offensive operations, providing the capability for targeted strikes with minimal collateral damage – at least theoretically. This capability to conduct precision strikes allows states to engage in targeted military action without a significant ground presence, limiting casualties and avoiding broader conflict escalations – a key factor motivating acquisition in many contexts.

The integration of drones into national security strategies further highlights the significance of their acquisition. Many states are now incorporating UAVs into their overall military doctrine, using them to enhance situational awareness, conduct targeted strikes, and support ground troops. This integration necessitates the development of specialized training programs, the establishment of supporting infrastructure, and the coordination of intelligence efforts to maximize the effectiveness of drone deployments. Furthermore, the increasing sophistication of drone technology necessitates a continuous effort to adapt national security doctrines to leverage these capabilities. This, in turn, further reinforces the drive to acquire the latest drone technology to maintain operational efficiency and strategic advantage.

The influence of major drone manufacturers is also a considerable factor. A small number of countries dominate the global drone market, supplying drones and related technologies to numerous states worldwide. These manufacturers play a significant role in shaping the global proliferation of drone technology, influencing the design, capabilities, and pricing of available systems. Their technological influence is profound, setting standards and influencing technological trajectories across the global market, effectively shaping the future of drone technology on a global scale. The relationship between these manufacturers and state actors is often complex, involving technological transfer, training, and even direct operational support. This intricate interplay between manufacturers and states deserves close scrutiny when analyzing the global proliferation of drone technology.

Regional case studies vividly illustrate the diverse ways in which states acquire and utilize drone technology. For example, in the Middle East, the proliferation of drones has contributed to ongoing conflicts, with multiple nations deploying drones for both offensive and defensive operations. The emergence of sophisticated indigenous drone programs in some Middle Eastern countries underscores the potential for rapid technological advancement and the need for careful monitoring of regional proliferation patterns. In other regions, such as Eastern Europe, the acquisition of drones reflects concerns about national security and territorial integrity, with a focus on surveillance and reconnaissance capabilities. The diverse geopolitical context of each region necessitates a tailored analysis to fully understand the implications of drone technology adoption.

The complexities surrounding drone acquisition highlight the need for a nuanced understanding of the interplay between technological advancements, geopolitical considerations, and the strategic choices made by individual states. The global distribution of drone technology, influenced by a mix of economic and military priorities, presents both opportunities and challenges for international security. The ease of acquisition coupled with the potential for misuse necessitates ongoing monitoring and the development of effective international cooperation mechanisms to address the potential risks posed by the uncontrolled proliferation of this transformative technology. A global dialogue encompassing technological innovation, ethical considerations, and legal frameworks is crucial to navigate the complexities of drone technology acquisition and deployment, ultimately promoting global stability and security rather than powering conflict and instability. The absence of a comprehensive, internationally recognized regulatory framework exacerbates these risks, emphasizing the urgency for concerted international action in this sphere. The proliferation of this technology highlights the need for greater transparency, international cooperation, and a clear understanding of the long-term implications for global security and the ethical use of this potentially destabilizing technology.

However, the picture of drone proliferation remains incomplete without examining the acquisition and utilization of unmanned aerial vehicles by non-state actors (NSAs). This presents a significantly different and, in many ways, more challenging aspect of the broader drone warfare landscape. Unlike state actors, NSAs often lack the resources, infrastructure, and regulatory frameworks needed for legal and transparent acquisition. Their methods are often clandestine, relying on a combination of illicit procurement, technological improvisation, and opportunistic acquisition.

The acquisition methods employed by NSAs are remarkably diverse and reflect their limited resources and the often-secretive nature of their operations. One common route is the purchase of commercially available off-the-shelf drones. These readily available systems, frequently marketed for civilian purposes such as photography and surveillance, offer NSAs a relatively easy path to acquiring basic drone capabilities. Many readily available drone models, with minor modifications, can be adapted for military purposes, carrying explosives or conducting surveillance operations. The ease of acquiring these drones online, coupled with the relatively simple modifications required to adapt them for military use, presents a significant challenge to control and regulation. The decentralized nature of online commerce makes tracking the flow of these drones incredibly difficult further compounding the problem.

Furthermore, NSAs increasingly leverage the black market to acquire more sophisticated drone technology, including components and fully assembled systems. This clandestine market allows them to bypass official channels and access potentially more advanced capabilities. These channels often involve intermediaries who facilitate the transfer of technology and equipment across borders, sometimes with the complicity of corrupt officials or individuals. The opacity of these transactions makes it exceedingly difficult to track the flow of weapons systems and to identify those responsible for supplying them. The lack of transparency also significantly hinders the ability of international organizations and law enforcement agencies to effectively interdict these illicit transactions.

Another method NSAs employ is the acquisition of drone technology through theft or capture. In conflict zones, the opportunity to seize drones from opposing forces, either through direct combat or through intelligence operations, presents a valuable avenue for NSAs to acquire advanced military capabilities. Similarly, theft from civilian or military facilities presents another route for obtaining drone technology, with captured drones often providing valuable technical insights and operational experience. This often leads to a rapid improvement in capabilities and the adaptation of seized technology.

Beyond the acquisition of complete systems, NSAs frequently demonstrate an impressive ability to adapt and modify existing drone technologies. This ingenuity often involves reverse engineering, utilizing readily available parts, and utilizing open-source information to improve their capabilities. This capacity for improvisation is amplified by the proliferation of open-source technical information and online communities where individuals share knowledge and techniques. This grassroots technological adaptation poses a significant challenge, as conventional arms control measures may be less effective against this rapidly evolving and often decentralized approach.

The implications of NSA drone acquisition for asymmetric warfare are profound and far-reaching. The ability of NSAs to acquire and effectively utilize drones fundamentally alters the balance of power in many conflicts. The use of drones by NSAs can significantly extend their reach, enabling them to conduct surveillance, reconnaissance, and even targeted attacks at minimal risk. Their ability to operate remotely, often at a significant distance, gives NSAs the capacity to strike high-value targets or military infrastructure while reducing the risk of direct engagement and casualties within their own ranks. This also enhances their capacity for propaganda and psychological operations by disseminating videos of attacks and showcasing their capabilities to a wider audience.

The types of operations conducted by NSAs using drones vary considerably based on their resources, technical expertise, and the specific conflict. Smaller, commercially available drones are frequently used for surveillance, gathering intelligence on military installations, troop movements, and civilian infrastructure. The information gathered enhances situational awareness, allowing for more effective planning of attacks or other operations. More sophisticated drones, when available, may be employed for targeted attacks, delivering small payloads of explosives or improvised explosive devices to inflict casualties and damage infrastructure. The precision capabilities of these attacks may be limited compared to state-actor deployments, but they still pose a serious threat.

Moreover, the use of drones by NSAs greatly complicates humanitarian efforts and international security. The increased frequency of attacks by drones, regardless of their sophistication, leads to civilian casualties and an escalation of conflict. Attribution of drone attacks can be extremely difficult, creating ambiguity and making it challenging to implement accountability mechanisms. This ambiguity can escalate tensions and lead to a cycle of retaliation and counter-retaliation. The decentralized nature of many NSA drone deployments also poses challenges to international cooperation, as traditional arms control mechanisms and international law may prove less effective in addressing these issues.

Addressing the proliferation of drones among NSAs requires a multi-faceted approach. Technological solutions, such as drone detection and jamming systems, are decisive but must be complemented by broader strategies. International cooperation is vital to establish clear regulations, share intelligence, and coordinate enforcement efforts. Supporting capacity building within vulnerable states to improve their internal security and regulate the drone market is also essential. This includes developing appropriate legislation, training law enforcement and security personnel, and enhancing national monitoring capabilities. Furthermore, focusing on countering the illicit financing of NSA acquisition and promoting greater transparency in the drone industry can also help disrupt the flow of drones into the hands of these groups. The absence of a robust international framework exacerbates the challenges posed by NSA drone acquisition, highlighting the urgent need for global collaboration to develop effective counter-measures and preventative strategies. The long-term impact of this proliferation on global security is significant and requires a sustained and collaborative global response.

Counter- Drone Technologies

Electronic warfare (EW) represents an important aspect of counter-drone strategies, focusing on disrupting the communication and navigation systems vital to drone operation. Jamming, a core component of EW, involves transmitting signals that interfere with the drone's ability to receive and process commands from its controller or to communicate with its supporting infrastructure. The effectiveness of jamming depends on various factors, including the type of jamming technology used, the drone's communication protocols, the distance between the jammer and the drone, and the environmental conditions.

Different jamming technologies exist, each with its strengths and weaknesses. Broadband jamming, for instance, covers a wide range of frequencies, disrupting multiple communication channels simultaneously. However, this approach consumes significant power and may have unintended consequences by affecting nearby legitimate communication systems. Narrowband jamming, conversely, targets specific frequencies used by a particular drone model, minimizing the impact on other devices. This method is more precise but requires detailed knowledge of the drone's communication system. Furthermore, sophisticated drones might employ frequency-hopping spread spectrum (FHSS) technology, making them more resilient to narrowband jamming. FHSS rapidly switches between frequencies, making it difficult for a single-frequency jammer to maintain consistent interference.

Directed energy weapons (DEWs), such as high-power microwave (HPM) systems, represent another evolving EW capability. These systems can disable or damage drone electronics directly, offering a more potent countermeasure than conventional jamming techniques. However, DEWs require advanced technology and can also have potentially harmful effects on other electronic systems and personnel.

The effectiveness of jamming against various drone platforms differs considerably. Simple, commercially available drones often rely on relatively unsophisticated communication protocols, making them vulnerable to straightforward jamming techniques. However, military-grade drones and those equipped with advanced anti-jamming features present a more significant challenge. These drones might employ multiple communication channels, redundant systems, and advanced signal processing algorithms to overcome jamming attempts. Furthermore, drones utilizing GPS augmentation or inertial navigation systems can continue functioning even when their primary communication link is disrupted.

Beyond the technical aspects, ethical considerations significantly impact the use of drone jamming. The indiscriminate nature of broadband jamming can disrupt vital civilian communications, including emergency services, air traffic control, and medical devices, causing significant harm. Narrowband jamming reduces this risk but still requires careful planning and execution to avoid unintended interference. The potential for collateral damage necessitates strict protocols and careful targeting to minimize harm to non-combatants and civilian infrastructure. Moreover, the legal and international frameworks governing EW operations are complex and constantly evolving. International humanitarian law (IHL) imposes strict limitations on the use of force, requiring proportionality and distinction between combatants and civilians. Jamming operations must comply with these principles, ensuring that the potential benefits outweigh the risks to non-combatants. National regulations also define the legal boundaries of EW operations, varying from country to country.

The development of effective counter-jamming techniques by drone manufacturers presents a continuous challenge to EW capabilities. Adaptive jamming, for example, dynamically adjusts jamming parameters to overcome anti-jamming measures, maintaining interference effectiveness. This technological arms race requires ongoing research and development efforts to maintain the superiority of counter-drone systems. This includes sophisticated signal processing algorithms to identify and classify drone signals, adaptive jamming techniques, and the development of new EW countermeasures tailored to specific drone models and communication protocols.

The integration of AI and machine learning into EW systems is rapidly changing the landscape. AI-powered EW systems can autonomously detect, identify, and neutralize drones with higher efficiency and accuracy compared to human-operated systems. Machine learning algorithms can learn and adapt to new drone communication protocols and anti-jamming techniques, providing an advantage over static jamming technologies. However, the increasing automation of EW operations raises ethical and safety concerns, highlighting the need for rigorous testing, verification, and clear guidelines for their deployment.

Furthermore, the use of electronic support measures (ESM) plays a critical role in effectively deploying jamming systems. ESM systems detect, locate, and identify the signals emitted by drones, providing essential intelligence for planning and executing EW operations. This information can be utilized to determine the appropriate jamming technique, optimize the jamming signal parameters, and predict the drone's potential manoeuvres. The combined use of ESM and jamming systems enhances the effectiveness of counter-drone measures.

The future of EW in counter-drone strategies involves a further integration of AI, improved sensor technology, and the development of more sophisticated jamming techniques. The development of advanced, adaptable jamming systems capable of handling multiple drone platforms and communication protocols simultaneously is likely to be a key area of development. Similarly, the exploration of novel jamming techniques, such as those employing artificial intelligence (AI) to dynamically adapt to evolving drone capabilities, will likely dominate the future EW landscape.

Beyond electronic warfare, the realm of anti-drone weaponry encompasses a diverse array of kinetic and non-kinetic systems, each with its own strengths, weaknesses, and strategic implications. Kinetic weapons, designed to physically destroy or disable drones, range from relatively simple, readily available options to sophisticated, high-tech systems. These include everything from small arms fire, shotguns, and even specialized anti-material rifles, effective against smaller, slower-moving drones, particularly those operating at close range. However, their accuracy and effectiveness diminish significantly against faster, more manoeuvrable drones, or those operating at higher altitudes. The use of kinetic weaponry also poses a substantial risk of collateral damage, particularly in densely populated areas. The potential for misidentification and accidental targeting of civilian aircraft or other assets is a critical concern. Moreover, the effectiveness of small arms fire against drones with protective casings or reinforced structures is notably limited.

Directed-energy weapons (DEWs), such as high-power lasers and microwaves, represent a more advanced approach to kinetic anti-drone technology. High-power lasers, for example, can precisely target and disable drones by overheating their electronics or causing mechanical damage. Their ability to engage targets at longer ranges and with pinpoint accuracy offers a significant advantage over conventional kinetic weapons. However, the technology is still in its relative early years, with high costs, limitations in atmospheric conditions, and potential eye-safety concerns significantly hindering widespread deployment. The power requirements for effective engagement are substantial, requiring large and powerful energy sources, often making them less mobile and easily deployable than other options. Furthermore, atmospheric conditions such as fog, rain, or dust can significantly reduce the range and effectiveness of laser systems.

High-power microwave (HPM) systems represent another type of DEW, capable of disabling drone electronics through electromagnetic pulses. These systems can disrupt a drone's communication, navigation, and control systems, rendering it inoperable. Unlike lasers, HPM systems are less susceptible to atmospheric interference, making them potentially more effective in a wider range of environmental conditions. However, HPM systems also have limitations, including the potential to damage or interfere with other electronic equipment within their range, raising significant safety concerns. Careful targeting and deployment protocols are essential to mitigate such risks. Furthermore, the development of drone systems with robust electromagnetic shielding presents a constant challenge to HPM technology, creating an ongoing technological arms race.

Beyond DEWs, missile systems offer a more powerful kinetic counter-drone solution. Small, lightweight missiles, often guided by infrared or radar, can effectively engage and destroy drones at longer ranges. These systems provide superior range and lethality compared to small arms fire, and they can better handle faster-moving and more agile drones. However, the cost per engagement is significantly higher than other options, and there is the risk of collateral damage should the missile malfunction or stray from its intended trajectory. The complexity of missile systems also requires more extensive training and maintenance than simpler, non-kinetic methods. The strategic implications of deploying missile systems for anti-drone operations are considerable, potentially escalating conflicts and raising concerns about the potential for misuse.

Non-kinetic anti-drone weapons represent a complementary approach, focusing on disabling drones without causing physical destruction. These systems offer several advantages, such as reduced risk of collateral damage and the potential for capturing the drone intact, which may provide valuable intelligence. Among the most common non-kinetic methods are nets and traps, designed to physically ensnare or capture a drone. These systems are often deployed in stationary positions, such as in airports or other sensitive areas, to prevent drone incursions. Nets are usually launched via pneumatic or spring-loaded mechanisms, and traps are typically designed to trigger upon drone contact, incapacitating the aircraft. Their effectiveness is limited against highly manoeuvrable drones, however, and their deployment requires careful planning and positioning.

Other non-kinetic methods include GPS jamming and radio frequency (RF) signal disruption. RF disruptors can interfere with a drone's communication link, preventing it from receiving control commands or transmitting data. Similar to jamming technology, this approach carries the risk of unintended consequences, particularly if operating in close proximity to legitimate RF communication systems. Furthermore, sophisticated drones may employ anti-jamming or redundant communication systems, rendering RF disruption ineffective. The development of effective countermeasures by drone manufacturers necessitates a continuous process of updating and improving these systems to maintain effectiveness.

The cost-effectiveness of different anti-drone systems varies considerably. Smaller, simpler systems, such as nets or specialized small arms, offer relatively low costs, but are limited in their range, accuracy, and effectiveness against sophisticated drones. DEWs and missile systems, on the other hand, are far more expensive but offer greater range, accuracy, and lethality, making them suitable for higher-value targets and more challenging scenarios. The choice of system depends heavily on operational requirements, budget constraints, and the specific threat posed by the drones being targeted.

The strategic implications of anti-drone weapons systems are far-reaching. The widespread availability of small, commercially available drones poses a significant threat to military installations, critical infrastructure, and civilian populations. The deployment of effective anti-drone technologies is therefore essential to mitigating this threat. However, the use of kinetic weapons, particularly missiles, can easily escalate conflicts, especially if employed against drones associated with state actors. The potential for accidental engagements or misidentification significantly increases the risk of escalation. Careful consideration of the proportionality of force and the potential for unintended consequences is fundamental when deploying any anti-drone weapons system.

The development of anti-drone weapons systems represents constantly evolving scenery. This ongoing technological arms race necessitates continuous research, development, and testing to maintain an effective defence against the diverse and rapidly evolving threat of unmanned aerial vehicles. The integration of artificial intelligence (AI) and machine learning (ML) offers a pathway to significant improvements in the speed, accuracy, and effectiveness of anti-drone systems. AI-powered systems could enable automated threat detection, classification, and neutralization, reducing reliance on human operators and potentially improving response times. The future of anti-drone warfare will undoubtedly be shaped by the integration of AI and ML, along with continued advancements in kinetic and non-kinetic technologies. However, these advances also needs careful consideration of the ethical and legal implications of increasingly autonomous weapons systems. The potential for unintended consequences, particularly in relation to civilian casualties, requires careful planning, thorough testing, and robust regulatory frameworks. The most common method of drone hacking involves exploiting vulnerabilities in the drone's communication protocols. Many drones rely on readily accessible and often unencrypted communication channels, such as Wi-Fi or Bluetooth, to transmit control signals and receive data from their operators. These open communication pathways offer an easy entry point for hackers to intercept and manipulate data streams. By injecting malicious code or intercepting control signals, an attacker can gain control of the drone, redirecting its flight path, manipulating its camera, or even disabling its functions entirely. Sophisticated attacks might involve injecting false GPS coordinates, causing the drone to deviate from its intended course and potentially crash or land in an unintended location. Similarly, an attacker could manipulate the drone's camera feed, inserting false images or obscuring the operator's view.

The sophistication of drone hacking techniques varies considerably. While less sophisticated attacks might involve simple signal jamming or spoofing, more advanced attacks leverage sophisticated software exploits to compromise the drone's onboard systems. These exploits often target vulnerabilities in the drone's firmware, the low-level software that controls its hardware. Firmware vulnerabilities can be exploited to gain complete control of the drone, allowing an attacker to perform a range of malicious actions. The development of sophisticated hacking tools and techniques has made it increasingly easier for malicious actors to compromise drone systems, even those with relatively robust security features.

One particularly insidious form of drone hacking involves the use of malware. Malware can be introduced into the drone's system through various methods, including infected software updates, compromised communication channels, or even physically implanted devices. Once inside the system, malware can perform various malicious functions, ranging from simple data theft to complete control of the drone's flight and sensor capabilities. Malware can also be designed to spread to other drones on the same network, creating a cascading effect and compromising a large number of devices simultaneously.

The potential consequences of drone hacking are significant. In a military context, a compromised drone could be used to gather sensitive intelligence, conduct surveillance, or even launch attacks against friendly forces. The potential for a drone to carry weapons and used against critical infrastructure, such as power grids or airports, is also a major concern. In civilian settings, the consequences could be equally severe, with the potential for privacy violations, property damage, and even loss of life. The ability to manipulate a drone's flight path or camera feed could be used to commit crimes, conduct reconnaissance, or even assist in terrorist activities.

To mitigate the risks associated with drone hacking, manufacturers and military operators are increasingly implementing robust cyber security measures. These measures include the encryption of communication channels, the use of secure firmware updates, and the implementation of intrusion detection systems. Encryption helps to protect data from interception, while secure updates ensure that the drone's software is free from vulnerabilities. Intrusion detection systems can monitor the drone's systems for signs of malicious activity, alerting operators to potential threats. Furthermore, rigorous testing and validation of drone software and hardware are important to identifying and mitigating vulnerabilities before they are exploited by malicious actors. Employing diverse authentication methods, including multi-factor authentication, can significantly enhance security.

The development of effective cyber security measures is an ongoing process. As drone technology continues to advance, so too must the sophistication of cyber security protocols. The development of new hacking techniques necessitates a continuous cycle of identifying vulnerabilities, implementing countermeasures, and testing the effectiveness of security protocols. This constant arms race between hackers and developers requires close collaboration between manufacturers, operators, and cyber security experts to stay ahead of the evolving threat.

The increasing integration of artificial intelligence (AI) into drone systems presents both opportunities and challenges in the realm of cyber security. AI can be used to enhance security measures, providing more sophisticated threat detection and response capabilities. However, AI-powered drones also present new vulnerabilities, as AI systems can themselves be subject to adversarial attacks. The development of AI-resistant cyber security measures is critical to protecting AI-powered drones from malicious actors.

The future of drone cyber security will likely be characterized by increasingly sophisticated and integrated security protocols. The use of block chain technology could provide enhanced security and transparency, ensuring the integrity of drone data and operations. Furthermore, the development of more robust and resilient drone systems, combined with enhanced cyber security training for operators and other personnel, will be central in mitigating the risks associated with drone hacking. International cooperation and the sharing of best practices are also essential to establish a global standard for drone security. This collaborative approach will be essential in fostering responsible development and deployment of this rapidly evolving technology. Failure to address these complexities could lead to widespread disruption, security breaches and potentially even the loss of life. The challenges of drone cyber security are significant, but proactive measures, combined with ongoing research and development, can mitigate the risks and ensure that the benefits of drone technology are realised while minimising potential harm.

Passive defence strategies focus on minimizing the vulnerability of assets and personnel to drone attacks. This primarily involves reducing the detect ability and attractiveness of targets. Camouflage and concealment techniques play a key role. This could range from employing camouflage netting over critical infrastructure to strategically positioning sensitive equipment within structures that offer natural shielding from aerial surveillance. The use of materials that absorb or deflect radar signals can further enhance concealment, making it more difficult for drones to detect and locate targets. Furthermore, limiting the visibility of sensitive activities and assets reduces the potential for drone-based reconnaissance and targeting. This could involve adjusting operational schedules, limiting illumination at night, or employing other measures to reduce the visual signature of the target.

Active defence measures involve taking direct action to neutralize or disrupt drone operations. A key aspect of this is the deployment of electronic countermeasures (ECM). These systems employ various techniques to interfere with the drone's communication links, GPS signals, or onboard control systems. Jamming technology disrupts the drone's ability to receive signals from its operator, effectively disabling its control. GPS spoofing can mislead the drone's navigation system, causing it to deviate from its intended course or land prematurely. Furthermore, radar systems can detect and track drones, providing early warning of potential threats and allowing for a timely response. The development and use of directed energy weapons (DEW), such as lasers, are also becoming increasingly relevant in counter-drone operations. These weapons can disable drones by targeting their sensors, communication systems, or propulsion systems, thereby eliminating the threat without causing collateral damage. However, the effectiveness of DEWs is highly dependent on factors such as weather conditions and the distance to the target.

Beyond individual technologies, the integration of multiple systems into a comprehensive counter-drone system is vital. A networked system combining radar detection, electronic countermeasures, and directed energy weapons, managed through a central command and control system, offers a layered defence approach that enhances effectiveness and situational awareness. Such a system allows the identification, tracking, and neutralization of drones across a wide area, providing a significant advantage in countering drone threats. Artificial intelligence (AI) plays an increasingly important role in the automation and optimization of these integrated systems. AI-powered algorithms can enhance threat identification, prioritize targets, and dynamically adjust countermeasures based on real-time data and threat assessments. This is particularly relevant in scenarios involving multiple drones or drone swarms.

Effective counter-drone strategies also necessitate significant organizational adjustments. Improved intelligence gathering and analysis are critical to anticipating and pre-empting drone attacks. This involves the monitoring of social media and open-source intelligence (OSINT) to identify potential threats, as well as utilizing human intelligence (HUMINT) and signals intelligence (SIGINT) to detect and track drone activity. Enhanced training and preparation for personnel are vital. This includes providing training on the identification of drones, the use of counter-drone equipment, and the development of standard operating procedures (SOPs) for responding to drone threats. Regular exercises and simulations should be implemented to test and refine operational responses and ensure that personnel are well-equipped to handle real-world situations. Furthermore, close cooperation and information sharing between different agencies and organizations are essential for a comprehensive and coordinated approach to counter-drone operations. This collaborative framework promotes a unified response, avoiding duplication of effort and leveraging the expertise of various stakeholders.

One significant challenge arises from the potential for drone swarms. These coordinated attacks, involving multiple drones working together, pose a particularly complex threat due to their ability to overwhelm existing defence systems. Countering drone swarms requires a shift in strategic thinking, focusing on disrupting communication and coordination rather than simply neutralizing individual drones. This necessitates the development and deployment of advanced jamming techniques that can disrupt the communication links between the drones and their operator. AI-powered systems can play a critical role in identifying and prioritizing targets within a swarm, optimizing the allocation of countermeasures, and managing the complexity of multiple simultaneous threats. The development of autonomous counter-drone systems that can independently identify, track, and neutralize multiple drones simultaneously is essential in overcoming the challenges presented by drone swarms.

The future of counter-drone technology will likely see increased sophistication and integration. Advanced sensor technologies, including AI-enhanced image recognition and radar systems, will enhance the ability to detect and identify drones at greater ranges and under diverse environmental conditions. More effective and precise DEW systems, capable of neutralizing drones at longer ranges and with greater accuracy, are also under development. The integration of counter-drone technologies with other security systems, such as perimeter security and cyber security measures, will create a more holistic and resilient approach to threat mitigation. Moreover, the continuous advancement of AI and machine learning will further refine the ability to predict, detect, and respond to drone threats in real-time. This evolution requires a commitment to ongoing research, development, and collaboration among governments, private sector companies, and academic institutions to stay ahead of the ever-evolving drone threat landscape.

The legal and ethical considerations surrounding counter-drone operations also require careful attention. The use of electronic countermeasures and directed energy weapons raises concerns about potential interference with civilian airspace and communications. Moreover, the deployment of autonomous counter-drone systems raises questions about accountability and the potential for unintended consequences. International cooperation and the development of clear guidelines and regulations are essential to ensuring the responsible and ethical use of counter-drone technologies, minimizing unintended harm to civilians and preserving the integrity of critical infrastructure. The establishment of clear protocols and regulatory frameworks will be essential to managing the deployment and operation of counter-drone systems effectively while ensuring compliance with national and international law. This complex interplay between technology, strategy, and legal frameworks will be fundamental in ensuring that the benefits of this vital technology are realized while minimizing the associated risks. The challenges ahead are significant, but through concerted effort, we can forge a path that balances the need for effective defense with the imperative to act ethically and responsibly.

One of the most significant developments will be the increased integration of artificial intelligence (AI) across all aspects of counter-drone systems. AI-powered algorithms are already enhancing threat identification, offering faster and more accurate detection of drones amidst background noise and clutter. This capability is further enhanced by advanced machine learning techniques that allow the system to learn and adapt to new drone types and tactics, constantly improving its detection accuracy and response time. Beyond detection, AI will play a central role in automated target prioritization and the dynamic allocation of countermeasures. In scenarios involving multiple drones or swarms,

AI will be vital in managing the complexity of simultaneous threats, optimizing resource allocation, and ensuring a coordinated and effective response. AI-driven predictive analytics can also assist in anticipating potential drone attacks by analyzing patterns of drone activity, identifying potential launch points, and predicting likely targets. This proactive approach to threat mitigation significantly improves overall security.

Furthermore, the development of autonomous counter-drone systems will be a defining characteristic of the future. These systems will be capable of independently identifying, tracking, and neutralizing drone threats without human intervention. This level of autonomy will be essential in scenarios where immediate action is required, such as in responding to fast-moving or unpredictable drone attacks. Autonomous systems will be capable of coordinating their actions with other counter-drone elements, creating a more robust and resilient defensive network. The level of autonomy can range from semi-autonomous systems that require human oversight to fully autonomous systems capable of operating independently. The implementation of such autonomous systems will however necessitate robust safety protocols and fail-safes to mitigate the risk of accidental harm to civilians or unintended consequences.

The sophistication of sensors will also significantly improve. Advanced radar systems, utilizing technologies such as phased array radar and synthetic aperture radar (SAR), will offer greater detection ranges and improved accuracy in identifying and tracking drones, even in challenging environments. The integration of multiple sensor modalities, combining radar, infrared, and electro-optical sensors, will provide a comprehensive picture of the threat landscape, enhancing the ability to discriminate between hostile drones and benign aircraft. Furthermore, the use of AI-enhanced image recognition will enable rapid and accurate identification of drone models and operational characteristics, providing important intelligence for developing effective countermeasures. High-resolution cameras coupled with AI-powered image analysis will allow for the identification of markings, serial numbers, and other unique features, facilitating post-incident investigations and attribution of drone attacks.

Directed energy weapons (DEW) will also undergo significant advancements. Lasers and high-powered microwaves are currently under development, offering the potential for precise and non-kinetic neutralization of drones. These systems offer several advantages over traditional kinetic weapons, including reduced collateral damage, pinpoint accuracy, and the ability to engage multiple targets simultaneously. Future DEW systems will likely achieve greater range, higher power output, and improved targeting accuracy, making them a more effective and versatile tool in counter-drone operations. The development of adaptive optics, capable of correcting for atmospheric distortions, will further enhance the effectiveness of lasers in challenging weather conditions. Furthermore, research into smaller, more portable DEW systems will allow for their deployment on a wider range of platforms, including smaller vehicles and even handheld devices.

The integration of counter-drone technologies with other security systems will also be crucial. This includes integrating counter-drone systems with perimeter security systems, such as fences, cameras, and intrusion detection sensors, to create a layered defence approach. Cyber security measures will also play an increasingly important role, safeguarding against hacking attempts and ensuring the integrity of counter-drone systems. The ability to share information and coordinate responses across different security systems will be vital in enhancing overall security and preventing successful drone attacks. Real-time data sharing and the integration of various sensor modalities will allow for a more comprehensive and efficient response to drone threats.

The potential for an arms race in counter-drone technology is a significant concern. As drone technology continues to advance, so too will the development of counter-drone systems, leading to a potentially escalating cycle of technological innovation and counter-innovation. This arms race could have significant implications for future conflicts, potentially raising the stakes and increasing the risk of escalation. The development and proliferation of increasingly sophisticated counter-drone technologies could exacerbate existing geopolitical tensions and create new security challenges. Therefore, proactive measures to control the proliferation of these technologies are necessary.

International cooperation will be vital in mitigating the risks associated with an arms race. The establishment of international norms and standards for the development and deployment of counter-drone technologies is essential. This could involve the creation of international treaties or agreements that regulate the use of these technologies, preventing their proliferation and ensuring their responsible use. Transparency and information sharing between nations can contribute to a better understanding of the capabilities and limitations of different systems, promoting responsible innovation and reducing the likelihood of unintended consequences. Furthermore, collaborative research and development initiatives could facilitate the development of more effective and safer counter-drone technologies, while minimizing the risks associated with an arms race.

The transformative potential of artificial intelligence (AI) and autonomous systems is reshaping the operational scenery of military drones, extending their capabilities far beyond the limitations of human control. This evolution is not merely incremental; it represents a fundamental shift in how drones are designed, deployed, and employed, impacting every facet of modern warfare. The integration of AI is empowering drones with unprecedented levels of autonomy, enabling them to perform complex tasks, adapt to dynamic environments, and make critical decisions in real-time, all without direct human intervention.

Directed energy weapons leverage focused beams of energy, such as lasers or high-power microwaves, to disable or destroy targets. Unlike conventional weaponry, DEWs offer several advantages in the context of drone defence. They can engage targets at the speed of light, providing instantaneous effects. This is particularly critical against fast-moving drones, which can be difficult to track and intercept using traditional methods. Furthermore, DEWs can be highly precise, minimizing collateral damage compared to explosive weapons. Their ability to target specific components, like the drone's electronics or propulsion system, offers the potential for a "soft kill," disabling the drone without causing widespread destruction. This selective engagement is particularly valuable in civilian contexts, where minimizing collateral damage is paramount. Finally, DEWs can be relatively inexpensive to operate compared to other anti-drone systems, as their ammunition is essentially energy, eliminating the need for costly munitions.

High-energy lasers, currently a leading type of DEW, operate by focusing intense beams of light onto a target, generating enough heat to damage or destroy critical components. Different laser types exhibit varying characteristics, affecting their effectiveness against different drone platforms. Solid-state lasers, for example, offer high power output and relatively good beam quality, making them suitable for engaging smaller, slower drones at close range. However, their power consumption can be significant, requiring substantial power generation capabilities. Chemical lasers, on the other hand, are capable of delivering much higher power, enabling them to engage larger, more robust drones at longer ranges. However, they require complex and hazardous chemical handling, posing logistical and safety challenges. Fiber lasers represent a promising middle ground, offering high power output with enhanced reliability and reduced maintenance needs, although their cost remains relatively high.

The effectiveness of laser-based DEWs depends on various factors, including the laser's power output, the target's material properties, atmospheric conditions, and the range to the target. Atmospheric effects like scattering and absorption can significantly attenuate the laser beam, reducing its effectiveness at longer ranges. Furthermore, the material composition of the drone's components greatly influences its susceptibility to laser damage. For instance, drones constructed using composite materials, such as carbon fiber reinforced polymers, are more resistant to laser damage compared to drones with metal components. This necessitates the development of laser systems with sufficient power and beam quality to overcome these limitations. Advances in beam control technologies, aiming to mitigate the effects of atmospheric turbulence and maintain accurate beam pointing, are crucial for extending the effective range of laser DEWs.

High-power microwaves (HPM) represent another promising class of DEWs for drone defense. HPM systems operate by emitting high-intensity electromagnetic pulses that can disrupt or disable electronic components within the drone. Unlike lasers, HPM systems do not rely on thermal effects for neutralization. Instead, they induce electromagnetic interference (EMI) that can overwhelm the drone's onboard electronics, causing malfunctions or complete system failure. This makes HPM weapons particularly effective against drones reliant on sophisticated avionics and communication systems. The advantage of HPM lies in their ability to target electronic systems, potentially disabling the drone without causing physical damage, leading to a non-destructive soft kill. This is crucial when operating in densely populated areas or near sensitive infrastructure.

The effectiveness of HPM systems depends on various factors, including the power and frequency of the emitted pulses, the drone's susceptibility to EMI, and the distance between the weapon and the target. Shielding and hardening of the drone's electronic components can significantly reduce the effectiveness of HPM attacks. Therefore, the development of advanced HPM systems with high power output and frequency agility is crucial to overcome this challenge. Further research is also necessary to understand the vulnerability of different drone architectures and electronic components to different HPM waveforms. This includes analyzing the effects of HPM radiation on different types of microcontrollers, sensors, and communication modules to design countermeasures specifically tailored to the most critical drone components.

Despite the potential of DEWs in drone defence, significant challenges remain. The development of high-power, compact, and cost-effective DEW systems is ongoing. High-power lasers, for example, require substantial power sources, which can limit their portability and deployment flexibility. Moreover, the effectiveness of DEWs is often dependent on factors like weather conditions, atmospheric attenuation, and the target's material properties. Adverse weather conditions, such as fog, rain, or snow, can reduce the range and accuracy of DEW systems. This necessitates the development of DEW systems that are resilient to these environmental factors, potentially through advancements in beam control, adaptive optics, or alternative energy delivery mechanisms.

The integration of DEWs into existing defence systems also poses challenges. The seamless integration of DEW systems with existing command, control, communication, computers, intelligence, surveillance, and reconnaissance (C4ISR) architectures requires advanced software and hardware solutions. The development of user-friendly interfaces and effective training programs is crucial for the successful deployment of DEWs by military and civilian operators. Furthermore, the potential for unintended consequences and the need for robust safety mechanisms must be addressed through rigorous testing and validation procedures. To ensure responsible development and deployment, extensive safety protocols are necessary, particularly for laser systems to protect both operators and bystanders from potential harm. The development of sophisticated safety mechanisms, such as automated target recognition and tracking systems, is imperative to minimize the risk of accidental engagements or misidentification of targets.

The legal and ethical implications of employing DEWs against drones also warrant careful consideration. International law and regulations concerning the use of force must be strictly adhered to, and appropriate safety protocols must be developed to prevent accidental injuries or damage to non-combatants. The development of clearly defined rules of engagement and the establishment of robust accountability mechanisms are essential to ensure responsible and ethical use of DEWs in drone defence.

Autonomous Navigation

One of the most significant advancements is in the realm of autonomous navigation. Traditional drones rely heavily on pre-programmed flight paths and GPS signals for navigation. This makes them vulnerable to GPS jamming, electronic warfare, and unpredictable environmental factors. AI-powered autonomous navigation systems, however, leverage a sophisticated fusion of sensor data, including GPS (when available), inertial measurement units (IMUs), radar, Lidar, and visual images, to create a robust and resilient navigational capability. These systems employ advanced algorithms for simultaneous localization and mapping (SLAM), enabling drones to build a real-time 3D map of their surroundings, even in GPS-denied environments. This capability is decisive for operations in contested airspace or complex urban environments, allowing drones to navigate autonomously through challenging terrain and avoid obstacles effectively. Furthermore, machine learning algorithms allow the systems to adapt to changing conditions, learning from past experiences to improve navigation accuracy and efficiency.

Beyond simple navigation, AI is revolutionizing target recognition and identification. Human operators currently shoulder the burden of analyzing sensor data to distinguish between targets and non-combatants. This process is time-consuming, prone to human error, and can be significantly hampered by poor visibility or challenging environmental conditions. AI-powered systems, equipped with advanced computer vision and deep learning algorithms, can process vast quantities of sensor data far more rapidly and accurately than humans, identifying targets with high precision. These systems can be trained on massive datasets of images and other sensor data, enabling them to recognize various types of targets, including vehicles, personnel, and infrastructure, even under adverse conditions. Furthermore, these AI systems can incorporate contextual information, such as geographic location and intelligence data, to improve target identification accuracy and reduce the risk of collateral damage.

The integration of AI also extends to the realm of autonomous decision-making. Currently, most military drone operations require significant human input, even for seemingly simple tasks. AI is enabling drones to make autonomous decisions about target engagement, route planning, and even self-preservation. For instance, an AI-powered drone might autonomously assess the risk of engaging a target, taking into account factors such as collateral damage risk, the presence of friendly forces, and the potential for enemy countermeasures. This level of autonomy is not merely about speeding up the decision-making process; it's about enabling drones to make more informed and effective decisions in rapidly evolving situations where human intervention may be delayed or impossible.

The development of such systems demands rigorous testing and validation to ensure their reliability and safety, emphasizing ethical considerations and minimizing unintended consequences.

The development of robust and reliable AI systems for military applications presents significant challenges. One primary difficulty is the complexity of AI algorithms themselves. Creating systems that can reliably operate in unpredictable and dynamic environments necessitates sophisticated algorithms capable of handling uncertainty, noise, and adversarial inputs. These algorithms must also be computationally efficient enough to run on the limited processing power and energy capacity of onboard drone systems. The development of specialized hardware and software optimized for AI applications is crucial for overcoming these limitations.

Furthermore, ensuring the safety and reliability of autonomous drone systems is vital. The potential for malfunctions, unintended consequences, and adversarial attacks demands rigorous testing and validation procedures. These procedures must cover a range of scenarios, including failures of individual components, unexpected environmental conditions, and sophisticated attempts to compromise the AI system. Formal methods and verification techniques are essential tools for ensuring the reliability and safety of these complex systems. Rigorous testing also needs to address ethical considerations and potential biases in the AI algorithms. These biases could lead to unfair or discriminatory outcomes.

The ethical implications of increasingly autonomous drones are a matter of significant debate. As drones become more capable of independent decision-making, concerns about accountability, transparency, and the potential for unintended consequences become increasingly prominent. The use of lethal autonomous weapons systems (LAWS), which can select and engage targets without human intervention, raises deep ethical questions about the role of humans in warfare and the potential for unintended escalation of conflict. Establishing international norms and regulations governing the development and deployment of autonomous weapons systems is vital to mitigate these risks. The ethical considerations necessitate thorough analysis and debate to find a balance between innovation and responsible use of these powerful technologies. The development and deployment of ethical frameworks for the use of autonomous systems is an ongoing and complex challenge.

The future of military drone operations is inextricably linked to the continued advancement of AI and autonomous systems. The integration of AI will not only enhance the capabilities of individual drones, but it will also transform the way drones are employed in military operations. Swarm technologies, coordinated groups of drones operating autonomously, promise to revolutionize warfare, creating new tactical possibilities and overwhelming adversaries through sheer numbers and coordinated action. These swarms will require advanced AI algorithms for coordination, communication, and collective decision-making, pushing the boundaries of AI capabilities.

However, the full realization of the potential of AI-powered drones depends on addressing the challenges related to algorithm complexity, system reliability, and ethical considerations. The successful integration of AI into military drone technology demands a multidisciplinary approach involving experts in AI, robotics, engineering, ethics, and international law. Only through careful consideration of both the technological and ethical dimensions can we harness the transformative power of AI while mitigating potential risks and ensuring its responsible use in the context of military operations. The future battlefield will be defined by the interplay of advanced autonomous systems, requiring careful consideration of their capabilities and limitations to prepare for the next generation of warfare. The strategic implications of these advancements are vast and far-reaching, influencing the balance of power and potentially reshaping the future of conflict.

The proliferation of military drones, powered by rapid technological advancements, has inevitably sparked a parallel arms race in counter-drone technologies. The battlefield of the future will not only be defined by the capabilities of drones themselves, but also by the effectiveness of systems designed to neutralize them. One of the primary approaches to countering drones involves electronic warfare (EW). This encompasses a wide range of techniques aimed at disrupting or disabling enemy drones by interfering with their communication and control systems. Jammers, for example, emit radio signals that overwhelm or interfere with the drone's control link, effectively disabling its ability to receive commands from its operator. These jamming systems can range from portable, handheld devices to sophisticated, high-powered systems capable of covering vast areas. However, the effectiveness of jamming is highly dependent on factors such as the frequency used by the drone, the power of the jammer, and the range of the jammer compared to the distance of the drone to its operator. Furthermore, sophisticated drones may employ frequency-hopping spread spectrum (FHSS) technology, making them more resistant to simple jamming techniques. Adaptive jamming techniques are increasingly common to combat these FHSS systems.

Beyond simple jamming, more advanced EW techniques employ spoofing, where false signals are transmitted to confuse or mislead the drone's navigation system, causing it to deviate from its intended course or even return to its launch point. This technique is particularly effective against drones relying on GPS for navigation, potentially forcing them into a controlled crash or simply rendering them ineffective. Spoofing technologies are constantly being developed, with improvements focusing on increasing the realism and reliability of the false signals, making them increasingly difficult to distinguish from legitimate signals. The use of advanced signal processing techniques and AI-powered algorithms is key in enhancing the effectiveness of spoofing.

Another significant counter-drone technology employs directed energy weapons (DEWs). These weapons use focused beams of energy, such as lasers or microwaves, to disable or destroy drones. Laser weapons, in particular, are proving increasingly effective against smaller drones, capable of instantaneously disabling their critical components such as cameras, sensors, or even their flight control systems. However, laser technology is still developing and the effectiveness of lasers depends on factors such as atmospheric conditions, the range to the target, the drone's materials, and the power of the laser. High-powered lasers are required to neutralize larger or more heavily armored drones, creating technical and logistical challenges. Moreover, laser systems may be vulnerable to countermeasures such as smoke screens or mirrors, highlighting the ongoing arms race between DEW development and countermeasures.

Microwave DEWs operate on a different principle, aiming to disable drone electronics through excessive heat generation. The effectiveness of microwave DEWs is also affected by factors including weather conditions and the target's distance. These systems are generally less precise than laser systems and potentially pose risks to nearby personnel or equipment. The development of microwave systems is still in its earlier stages compared to laser systems, however, they do possess potential advantages against more heavily armoured drones or those employing countermeasures against laser-based systems.

Beyond DEWs, kinetic countermeasures represent another major category of counter-drone technology. This includes the use of conventional weapons like anti-aircraft guns or missiles, adapted for engaging smaller, faster-moving targets. Small caliber anti-aircraft weapons, equipped with advanced targeting systems, are effective against smaller drones. However, the use of these systems in populated areas poses significant risks. More specialized anti-drone missiles are being developed, offering greater precision and reduced collateral damage. These missiles, often utilizing proximity fuses or other advanced guidance mechanisms, can track and intercept drones with high accuracy, particularly when employing advanced image recognition to distinguish the drone from other objects. The development of these missiles is constantly evolving, focusing on increasing range, accuracy, and effectiveness against sophisticated drone evasion tactics.

The effectiveness of kinetic countermeasures is highly dependent on the ability to accurately track and identify the target drone amidst other aerial traffic. This needs the use of advanced radar systems and sensor fusion techniques, combining data from multiple sources to provide a comprehensive picture of the airspace. Advanced AI-powered algorithms are essential in filtering out noise, identifying potential threats, and guiding weapons to their targets. The development of sophisticated tracking systems is a key focus of counter-drone research, aiming to improve the probability of accurate engagement, reduce collateral damage, and neutralize drones before they can accomplish their mission.

Furthermore, the development of net-based countermeasures is gaining traction. These systems employ lightweight nets launched from either ground-based launchers or drones to physically capture drones in flight. This method is particularly effective against smaller drones and offers a non-destructive approach to neutralizing a threat, reducing the risk of collateral damage or causing harmful debris. However, these systems are limited in range and effectiveness against faster or more agile drones. The technology continues to evolve to enhance the range and accuracy of the nets as well as increase their resilience to evasive maneuvers.

A critical aspect of the counter-drone arms race involves the development of sophisticated cyber warfare techniques to disrupt or control enemy drones. This can involve hacking into the drone's control systems, potentially taking over its navigation or weapon systems. Cyber attacks can also be used to remotely disable or reprogram the drone's software, effectively rendering it useless. However, this approach is highly dependent on the security of the drone's software and its ability to withstand cyber attacks. The development of robust cyber security measures for drones is a critical area of ongoing research, aiming to make them more resistant to cyber intrusions.

The counter-drone scene is a dynamic and evolving environment. The arms race between drone technology and counter-drone capabilities is likely to continue, driving innovation and leading to the development of even more sophisticated systems. The effectiveness of any given counter-drone technology is highly dependent on various factors, including the specific type of drone being countered, environmental conditions, and the skill of the operators. A layered defence approach, combining multiple counter-drone technologies, is therefore likely to be the most effective strategy in neutralizing enemy drones. The integration of artificial intelligence and machine learning will play a significant role in optimizing counter-drone systems, enabling them to adapt to new threats and improve their effectiveness over time. As drones continue to evolve, so will the countermeasures designed to defeat them, perpetuating this vital aspect of modern military technological development. The future of warfare will significantly depend upon the ability to effectively leverage both drone technology and the means to counter it.

The development of effective counter-drone technologies is a complex and rapidly evolving field. The initial responses were often reactive, focusing on readily available technologies adapted for defensive purposes. Jamming devices, for example, were initially used to disrupt the communication links between drones and their operators, effectively disabling their functionality. However, the sophistication of drone technology quickly outpaced these early countermeasures. Modern drones are increasingly equipped with redundant communication systems and advanced anti-jamming techniques, rendering simple jamming ineffective against determined adversaries. This has necessitated the development of more sophisticated counter-drone systems.

One significant area of development is directed energy weapons (DEWs), such as high-powered microwave (HPM) systems. These weapons use focused beams of microwave energy to disable drones by disrupting their electronic components and rendering them inoperable. The advantage of DEWs lies in their ability to target and disable drones without causing physical damage, reducing the risk of collateral damage. However, the effective range and power of DEWs are still under development, limiting their current widespread applicability. Furthermore, the cost and complexity of deploying such systems are substantial, placing them beyond the reach of many nations.

Another significant countermeasure involves the development of specialized anti-drone weapons. These weapons range from small arms, such as shotguns and rifles, specifically designed to shoot down small drones, to more sophisticated systems like anti-aircraft cannons and missile systems adapted for use against drones. The effectiveness of these weapons depends heavily on factors such as the size and speed of the drone, the altitude of flight, and the range of the weapon system. Larger drones, particularly those with heavier payloads, present a more challenging target, demanding more powerful and sophisticated weapon systems. The integration of advanced sensor systems and target acquisition technologies has also been essential in improving the accuracy and effectiveness of anti-drone weaponry. However, the use of such weapons in civilian airspace poses significant risks of collateral damage and requires careful consideration.

Beyond the development of kinetic weapons, significant investments are being made in cyber security measures to protect against drone attacks. This involves developing secure communication protocols, robust software systems, and advanced encryption technologies to prevent unauthorized access and control of drones. This focus is particularly important given the growing use of drones for espionage, data collection, and other hostile activities. Advanced cyber security measures aim not only to protect against direct attacks but also to detect and mitigate potential vulnerabilities in drone systems before they can be exploited by adversaries. The development of artificial intelligence and machine learning is also playing an increasingly significant role, enabling the creation of systems capable of identifying and neutralizing potential drone threats autonomously. These systems can analyze vast amounts of data, identifying patterns and anomalies that may indicate hostile drone activity.

The arms race in drone technology is not solely focused on the development of counter-drone measures. Drone manufacturers and military organizations are continuously improving drone capabilities in response to the emergence of new countermeasures. This has led to a continuous cycle of innovation, with drones becoming increasingly difficult to detect and disable.

New Technologies

The integration of advanced sensor technologies is fundamentally altering the capabilities of military drones, significantly enhancing their situational awareness and effectiveness across a wide spectrum of missions. One particularly transformative technology is hyper spectral imaging, a technique that provides far richer information compared to traditional multispectral or colour images. While standard cameras capture images in a limited number of wavelengths (e.g., red, green, blue), hyper spectral sensors capture hundreds or even thousands of narrow, contiguous spectral bands across the electromagnetic spectrum. This capability unlocks unprecedented levels of detail about the composition and characteristics of objects and materials.

In the context of military applications, hyper spectral imaging offers several critical advantages. Consider the challenge of detecting camouflaged targets or concealed explosives. Human observers, even with enhanced vision equipment, may struggle to distinguish these from their surroundings. Hyper spectral sensors, however, can penetrate camouflage and detect subtle spectral signatures that reveal the underlying materials. This is because different materials absorb and reflect light differently at various wavelengths. A hyper spectral image reveals these subtle differences as unique spectral "fingerprints," allowing for precise identification of targets otherwise difficult or impossible to detect. For instance, a sniper hidden in foliage might be readily identified by the unique spectral signature of their clothing or equipment, even if visually camouflaged. Similarly, landmines or improvised explosive devices (IEDs) buried underground can often be detected through their unique spectral interactions with the soil, providing central pre-emptive information for clearing operations.

The implementation of hyper spectral sensors in drones offers a powerful combination of mobility and analytical capability. A drone equipped with such a sensor can survey a large area quickly, collecting detailed spectral data that is then processed by sophisticated algorithms to identify and classify targets. This is a significant improvement over ground-based systems, which are less mobile and have a more limited field of view. The ability to survey large areas rapidly is crucial for tasks like border security, search and rescue operations, or detecting enemy movements in a contested area.

The data generated by hyper spectral sensors is often very large and complex, requiring sophisticated algorithms for processing and analysis. Advanced signal processing techniques are used to extract meaningful information from the raw spectral data, and machine learning algorithms are increasingly employed to train models for automatic target recognition and classification. This automated analysis significantly reduces the workload on human analysts, allowing them to focus on higher-level decision-making and interpreting the overall intelligence picture. The development of robust, lightweight, and power-efficient hyper spectral sensors suitable for integration into drone platforms is an area of active research and development. Miniaturization and improved processing capabilities are critical for extending the operational range and endurance of these systems.

Beyond hyper spectral imaging, other advanced sensor technologies are being integrated into military drones to further enhance situational awareness. LiDAR (Light Detection and Ranging) systems are becoming increasingly prevalent, offering detailed 3D maps of the terrain, even in low-light or adverse weather conditions. LiDAR uses laser pulses to measure distances to objects, creating point clouds that can be used to reconstruct highly accurate 3D models. This is particularly valuable for navigation in complex environments, obstacle avoidance, and creating detailed battlefield maps for planning purposes. Combined with hyperspectral data, LiDAR can provide even more detailed information about the environment, including the three-dimensional structure and composition of buildings and other infrastructure.

Synthetic Aperture Radar (SAR) is another technology increasingly incorporated into military drones. SAR can operate regardless of weather conditions or time of day, making it ideal for all-weather reconnaissance. Unlike optical sensors, SAR actively transmits radio waves and receives the reflected signals, creating images based on the backscattered energy. Different materials have unique radar signatures, allowing SAR to distinguish between various objects and terrain features. This is particularly valuable in detecting camouflaged targets or identifying infrastructure in areas with limited visibility. SAR data can be integrated with other sensor data to create a complete and robust picture of the environment.

Thermal imaging, another important sensing technology, detects infrared radiation emitted by objects. This allows drones to detect temperature differences, which can be used to identify targets, even in complete darkness or through obscurants like smoke or fog. Thermal imaging is particularly useful for detecting human presence or identifying operating vehicles by their heat signatures. The ability to detect human body heat is highly valuable in search and rescue operations, while identifying the heat signatures of operating vehicles offers critical intelligence for targeting or surveillance purposes.

The fusion of data from multiple sensor types – hyper spectral, LiDAR, SAR, thermal, and even traditional visible-light cameras – is central for maximizing the effectiveness of drone-based intelligence gathering. Data fusion techniques combine information from different sensors to create a more comprehensive and accurate understanding of the environment and targets. This reduces ambiguity and improves the reliability of intelligence assessments. Advanced algorithms are used to correlate and integrate information from various sources, creating a holistic view that is greater than the sum of its parts. These fused datasets can be used to create detailed 3D models of the battlefield, allowing for more effective planning and execution of military operations.

Furthermore, the development of more sophisticated algorithms for processing and interpreting sensor data is a key driver in the advancement of drone capabilities. Machine learning and artificial intelligence are transforming the way sensor data is analyzed, enabling faster and more accurate target identification, tracking, and classification. The increasing sophistication of these algorithms is reducing the reliance on human analysts, enabling drones to perform increasingly autonomous and complex tasks. AI-powered systems can not only identify targets but can also predict their future behavior, allowing for preemptive action and improved operational effectiveness.

The miniaturization and increased efficiency of power systems are also enabling more advanced sensor integration. Smaller and lighter sensors consume less power, allowing drones to stay airborne for longer periods and carry a greater payload of sensors. The development of more efficient batteries and solar-powered systems is pushing the boundaries of drone endurance, extending their operational range and capabilities. This improved efficiency is critical for extended surveillance missions, where continuous monitoring is required over large areas.

The strategic implications of these technological advancements are far-reaching. The enhanced situational awareness afforded by advanced sensor technologies such as hyper spectral imaging, LiDAR, SAR, and thermal imaging is transforming the way modern militaries conduct reconnaissance, surveillance, and targeting. The ability to detect camouflaged targets, identify hidden explosives, and create detailed 3D maps of the battlefield provides a significant advantage in modern conflicts. The increasing autonomy of these systems, driven by AI and sophisticated algorithms, allows for faster decision-making and reduced reliance on human intervention, making military operations more efficient and effective.

However, these advancements also present challenges. The increasing complexity of sensor systems and the volume of data they generate necessitate highly sophisticated processing and analytical capabilities. The development and deployment of robust and reliable AI systems for autonomous target recognition and engagement raises ethical concerns about accountability and the potential for unintended consequences. International discussions and regulations regarding the use of autonomous weapons systems are essential to mitigating potential risks and ensuring responsible use of these powerful technologies. Further research and development are necessary to balance the advancements in technology with ethical considerations and international norms. The future of military drones hinges on a careful balance between innovation and responsible application, safeguarding against unintended consequences while maximizing the strategic advantages these technological leaps offer.

The proliferation of drone technology has fundamentally altered the geopolitical landscape, introducing both opportunities and significant challenges to international security. The accessibility and relative affordability of various drone platforms, particularly smaller, commercially available models that can be easily modified for military purposes, have democratized aerial warfare to a certain extent. This has led to a complex interplay of power dynamics, regional conflicts, and the potential for unforeseen consequences on a global scale.

One of the most immediate impacts is the shift in the balance of power. Previously, air power was largely the domain of major military powers with significant resources invested in advanced fighter jets and sophisticated air defence systems. The emergence of relatively inexpensive and readily deployable drones, however, has levelled the playing field somewhat, allowing smaller nations and even non-state actors to project military power beyond their traditional capabilities. This is particularly evident in asymmetric conflicts where smaller, less powerful entities can utilize drones for surveillance, reconnaissance, and even targeted attacks against larger, more technologically advanced adversaries. The effectiveness of this tactic was demonstrated in numerous conflicts, where drones played a significant role in changing the dynamics of battles.

Stealth technology, is being incorporated into drone design to reduce their radar signature and make them harder to track. Advanced materials and propulsion systems are also being developed to improve their speed, manoeuvrability, and endurance. The use of artificial intelligence and autonomous flight systems allows drones to operate independently, making them more difficult to target and control. This dynamic interplay between offensive and defensive technologies is driving a rapid pace of technological development, escalating the arms race and raising concerns about its long-term implications for international security.

The escalating arms race necessitates a concerted effort to mitigate its destabilizing effects. International arms control agreements play a crucial role in regulating the proliferation of drone technology and preventing an uncontrolled escalation. However, the lack of a universally agreed-upon definition of drones, coupled with the rapid technological advancements, makes it incredibly challenging to negotiate and enforce these agreements. The diverse range of uses for drones, from civilian applications to military operations, further complicates the discussion. Moreover, the difficulty in verifying compliance with arms control agreements adds to the challenge. The proliferation of drones among non-state actors, who often operate outside of established legal frameworks, presents an additional significant obstacle to effective arms control.

Proliferation of Drones

Addressing the proliferation of drone technology requires a comprehensive approach combining technological solutions with international cooperation and diplomatic efforts. Technological advancements in counter-drone technologies are essential but must be coupled with international efforts to develop clear guidelines and regulations for the development, production, and deployment of drones. Information sharing among nations is crucial for developing effective counter-measures and deterring potential threats. The establishment of international norms and standards for responsible drone usage is essential to mitigate the risk of escalation and minimize unintended consequences. The integration of technological capabilities and diplomatic initiatives is pivotal in addressing the complexities of the drone arms race, while encouraging responsible innovation and preventing a future of widespread drone conflict.

The ease of acquisition, coupled with the increasingly sophisticated capabilities of these unmanned aerial vehicles (UAVs), necessitates a robust international response aimed at regulating their production, deployment, and use. However, creating effective arms control mechanisms in this rapidly evolving technological landscape proves incredibly difficult. The inherent complexities of defining "drone technology," the diverse applications of UAVs, and the challenges of verification and enforcement all contribute to the obstacles faced in establishing meaningful international agreements.

One of the primary hurdles is the lack of a universally accepted definition of a "drone." The technological spectrum is vast, ranging from small, commercially available quadcopters to large, military-grade UAVs capable of carrying substantial payloads. This lack of a clear definition makes it challenging to establish clear criteria for regulation and control. An agreement that encompasses all types of UAVs risks being overly broad, potentially hindering legitimate civilian applications, while a narrowly defined agreement could leave critical loopholes that malicious actors could exploit. The rapid pace of technological advancement further exacerbates this problem. New features and capabilities are constantly emerging, making it difficult for international agreements to keep pace. Agreements that are painstakingly negotiated may quickly become obsolete, necessitating constant updates and revisions. This dynamic necessitates a flexible and adaptable framework for regulation, rather than a rigid set of rules that quickly become outdated.

Several initiatives and treaties have attempted to address specific aspects of drone proliferation. These efforts, while commendable, often lack the scope and universality needed to effectively tackle the challenge. The focus on certain types of drones or specific aspects of their operation often leaves gaps that malicious actors can exploit. Furthermore, the enforcement mechanisms associated with these initiatives frequently prove inadequate, hindering their effectiveness. International cooperation is paramount in enhancing the effectiveness of existing treaties and developing new frameworks for arms control. This includes information sharing, joint research initiatives, and collaborative capacity-building programs among countries. Capacity building plays a critical role in assisting developing nations in establishing regulatory frameworks, monitoring compliance, and countering threats from drones.

Despite the significant challenges, international cooperation and arms control efforts in the sphere of drone technology are showing some signs of progress. There's a growing recognition of the need for collective action to mitigate the risks associated with drone proliferation. However, the effectiveness of these efforts hinges on the commitment of states to participate in a meaningful and comprehensive manner. The absence of key actors from such negotiations weakens the authority and effectiveness of any agreements reached. The political dynamics at play, with varying national interests and priorities, further complicate the process. Negotiations must strike a balance between the legitimate security needs of states and the concerns for humanitarian law and civilian protection.

The proliferation of drone technology among non-state actors presents an additional challenge. These actors often operate outside of established legal frameworks and are less susceptible to international pressure. Addressing this issue requires a multi-faceted approach that combines law enforcement, intelligence gathering, and international cooperation to disrupt the supply chains and networks supporting the acquisition and use of drones by non-state actors. This necessitates a global effort involving information sharing, capacity building, and joint operations to dismantle these networks.

The future of international cooperation and arms control regarding drone technology remains uncertain. The rapid technological advancements and the evolving nature of conflict demand a flexible and adaptable framework for regulation. However, progress is dependent on the willingness of states to engage in meaningful negotiations and commitments to compliance. International organizations can play a vital role in fostering dialogue, facilitating information sharing, and promoting the development of common standards. However, ultimately, the success of international efforts hinges on the collective commitment of the international community to address the challenges posed by drone proliferation and to ensure responsible innovation and development in this critical technological domain. Without concerted and collaborative efforts, the unchecked proliferation of drones will continue to pose significant risks to global security and stability, demanding a continuous evolution of strategies, technologies, and diplomacy.

The accessibility of drone technology is rapidly shifting the global security scene. No longer are the exclusive domain of militaries and well-funded intelligence agencies, drones becoming increasingly affordable and readily available to a wider range of actors, including non-state actors, criminal organizations, and even individuals. This democratization of drone technology presents both opportunities and significant challenges. The decreasing cost of components, coupled with the proliferation of readily available DIY kits and online resources, lowers the barrier to entry for potential users, vastly expanding the potential pool of drone operators. This trend is particularly concerning given the potential for misuse, ranging from illicit surveillance and smuggling to targeted attacks and acts of terrorism.

The miniaturization of drone technology is another significant trend. Smaller, lighter drones are easier to conceal, transport, and operate discreetly, making them far more difficult to detect and intercept. These smaller systems can be deployed in confined spaces, such as urban environments or dense forests, where larger drones might be restricted. Their compact size also increases their portability, facilitating their use in remote or otherwise inaccessible locations. While smaller size might seem to reduce the potential for destructive impact, the miniaturization trend is accompanied by improvements in payload capacity and technological sophistication, potentially negating the supposed limitations of size. For example, micro-drones equipped with sophisticated cameras and high-resolution sensors can still gather valuable intelligence, while those carrying small explosive charges can inflict significant damage. This combination of size, portability, and capability makes smaller drones a particularly worrying development in terms of proliferation.

Moreover, the evolution of drone capabilities is dramatically changing the battlefield. The integration of advanced technologies, such as artificial intelligence (AI), machine learning, and swarm technology, enhances drones' effectiveness and autonomy. Swarms of small, inexpensive drones, each individually equipped with limited capabilities, can collaboratively achieve complex objectives that would be challenging, if not impossible, for a single, larger drone. These swarms could overwhelm defences through sheer numbers and coordinated actions, presenting a significant tactical advantage. Furthermore, the ability to autonomously perform complex tasks like target identification and engagement reduces the need for human intervention potentially increases the frequency and scale of drone operations. This reduces the risks faced by operators, making drone operations even more appealing.

The integration of artificial intelligence (AI) significantly amplifies the potential for both beneficial and detrimental uses of drones. AI-powered drones possess enhanced capabilities in autonomous navigation, target recognition, and decision-making. This autonomy can improve efficiency, reduce human error, and expand the range of tasks drones can undertake. However, the increasing autonomy raises serious ethical and safety concerns. The development of lethal autonomous weapons systems (LAWS), commonly known as "killer robots," represents a particularly sensitive aspect of AI integration. The delegation of life-or-death decisions to machines raises profound ethical questions and necessitates careful consideration of the implications for international humanitarian law and human rights.

The emergence of swarm technology presents further complexities in the proliferation of drones. The ability to deploy a large number of small drones working in coordination poses unique challenges to existing defensive systems. These swarms can overwhelm defences through coordinated attacks, making it difficult to intercept or neutralize all drones within the swarm. The decentralized nature of swarm operations makes it harder to identify and target the source of the attack, making them a potentially potent weapon for both state and non-state actors. This capability could significantly alter the dynamics of future conflicts, demanding new approaches to defence and security.

The challenges of controlling the spread of drone technology are substantial. International arms control treaties have struggled to keep pace with the rapid advancements and diversification of drones. The lack of a universally accepted definition of a "drone" hampers the creation of effective regulations. The diverse applications of drones, ranging from civilian uses to military applications, make it challenging to establish clear guidelines without hindering legitimate activities. Furthermore, enforcing regulations internationally is difficult given the diverse levels of technological capacity and regulatory frameworks across countries.

Moreover, the proliferation of drone technology poses significant challenges to international security. The ease of acquisition, coupled with the advanced capabilities of these systems, increases the likelihood of their use in illicit activities and conflicts. The lack of effective international cooperation and regulation exacerbates the risks. Addressing this requires a multifaceted approach involving international collaboration, technological advancements in counter-drone systems, and robust regulatory frameworks. This should include strengthening export controls, promoting responsible innovation, and fostering international cooperation to share best practices in countering the threat posed by drone proliferation.

The future trajectory of drone technology is uncertain, but several trends are likely to continue. The cost of production is likely to decrease further, making drones even more accessible. The capabilities of drones will continue to advance, driven by ongoing improvements in AI, automation, and sensor technology. The development and deployment of counter-drone systems will become increasingly important to mitigate the risks associated with the proliferation of drones. These systems could range from electronic jamming and spoofing techniques to physical interception and destruction of drones. However, these countermeasures also face a constant technological arms race with the continuous improvement of the offensive capabilities of drones.

The proliferation of drone technology requires a global effort to address the associated challenges. This necessitates international cooperation to establish clear norms, standards, and regulations for responsible drone use. A collaborative approach involving governments, industry, and academia is critical to fostering innovation while mitigating the risks associated with the increasing accessibility and capabilities of drones. The development of effective counter-drone technologies, the enhancement of international intelligence sharing mechanisms, and the promotion of responsible AI development are essential to address these concerns. Ignoring the ongoing proliferation of drone technology is not an option. A proactive, collaborative, and comprehensive approach is imperative to navigate the complex landscape of drone technology and safeguard international security. The future of warfare, and indeed global stability, depends, in part, upon our ability to manage this potent and increasingly accessible technology.

Socioeconomic Impacts of Drone Warfare

The proliferation of drone technology has profound and multifaceted economic consequences, impacting both the nations involved in their production and deployment, and the regions where they are used in conflict. The economic impact extends beyond the immediate costs of manufacturing and military operations, encompassing broader effects on labour markets, research and development investments, and the overall economic health of affected regions.

One of the most significant economic impacts is the creation of jobs within the drone industry itself. This growing sector encompasses a wide range of activities, from the design and manufacturing of airframes and components to the development of sophisticated software, sensor technologies, and artificial intelligence algorithms. The manufacturing process generates employment in various skill levels, from assembly line workers to highly specialized engineers and software developers. The operational side of drone technology also creates jobs, including pilots, maintenance technicians, analysts interpreting drone imagery, and personnel responsible for managing drone fleets and coordinating missions. This substantial job creation is particularly significant in countries that have established themselves as major drone manufacturers and exporters, boosting their national economies and contributing to overall employment growth. The economic stimulus extends further to related industries, such as those supplying raw materials, components, and specialized services to the drone industry.

Furthermore, significant investments are made in research and development (R&D) to advance drone technology. Governments and private companies invest heavily in improving the capabilities of drones, encompassing advancements in autonomous flight, sensor technology, artificial intelligence, and payload delivery systems. This sustained R&D effort generates innovations that have broader economic applications, extending beyond the military sphere. The technologies developed for military drones often find applications in civilian sectors, such as agriculture, infrastructure inspection, search and rescue, environmental monitoring, and disaster relief. This technology transfer stimulates innovation and economic growth across various sectors, contributing to a broader economic ecosystem and fostering technological advancements with wider societal benefits.

The economic benefits of drone manufacturing and export are substantial for certain nations. Countries that have successfully established a strong drone industry benefit from export revenue, strengthening their balance of payments and contributing to their overall economic competitiveness. The export of drones, components, and related technologies can significantly enhance a nation's international trade position and boost its overall economic standing. These economic gains, however, are not evenly distributed, and are often concentrated in a relatively small number of technologically advanced countries. This disparity creates economic imbalances and can exacerbate geopolitical tensions, particularly between developed and developing nations. A nation's success in the drone sector requires a sophisticated technological infrastructure, skilled workforce, and significant investment in R&D, creating barriers to entry for many less developed countries.

on the contrary, the economic impact of drone warfare on affected regions is often devastating. The use of drones in conflict leads to the destruction of infrastructure, including roads, bridges, buildings, and critical utilities. The costs of repairing or replacing this damaged infrastructure can be substantial, straining the already limited resources of conflict-affected areas and hindering their economic recovery. Moreover, drone strikes often result in civilian casualties, resulting in significant economic losses through the loss of productive workforce members and the long-term costs of healthcare and social support for the injured and families of the deceased. The economic ripple effect extends to disruptions in trade, agriculture, and tourism, further hindering economic development and exacerbating existing inequalities. Businesses may be forced to close due to damage, loss of life, or disruption of supply chains, resulting in job losses and reduced economic activity. Furthermore, the psychological trauma inflicted by drone warfare often leads to reduced productivity and increased healthcare costs, representing a significant long-term economic burden. The reconstruction and rehabilitation efforts following conflict often demand significant international aid and investment, posing further economic challenges for the affected countries.

The economic costs associated with the use of drones in military operations, including the costs of R&D, deployment, and maintenance, are significant. Furthermore, the long-term consequences of drone warfare on affected regions, such as the ongoing instability and economic hardship in certain parts of the

Middle East and Afghanistan, are also substantial and represent a considerable indirect economic cost to the global community. Equally, countries like Israel, a major player in the development of drone technology, have leveraged their advancements in drone technology to secure lucrative export contracts, contributing to their economic growth. Yet, even in such cases, the ethical and economic implications of their drone deployments remain a subject of ongoing debate.

The economic impact of drone warfare also extends to the costs of counter-drone technologies. As drone technology continues to advance, so too does the development of countermeasures, necessitating substantial investments in defence systems, creating a cycle of escalating costs and an arms race. This economic competition for military technological superiority diverts resources from other vital sectors, such as healthcare, education, and infrastructure development, creating opportunity costs that impact a nation's overall welfare. These ongoing investments in counter-drone technologies represent a significant economic weight, further stressing national budgets and potentially hindering economic development in other areas. The economic implications of this technological competition are far-reaching and demand careful consideration of the trade-offs involved.

The economic consequences of drone warfare are only one facet of its broader societal impact. The pervasive use of drones in military operations has triggered a cascade of social and political ramifications that extend far beyond the battlefield, impacting public perception, international relations, and the very fabric of democratic governance. The lack of transparency surrounding drone programs, coupled with the often-unclear criteria for targeting, has led to a significant erosion of public trust in government institutions. Citizens, particularly in countries where drone strikes are frequent, feel increasingly alienated from their governments, fostering a climate of suspicion and mistrust. This diminished public faith undermines the legitimacy of political processes and weakens democratic accountability.

The opacity surrounding drone programs is a key contributor to this erosion of trust. Governments often cite national security concerns to justify the secrecy surrounding drone operations, making it exceedingly difficult for the public to assess the legality, morality, and effectiveness of these strikes. The lack of independent oversight and the limited avenues for redress for victims of drone strikes further exacerbate this issue. This lack of transparency allows for a perception of unchecked power and arbitrary decision-making, fuelling public cynicism and resentment. The public's demand for accountability clashes with the government's assertion of national security, creating a persistent tension that erodes the fundamental principles of open and transparent governance.

Furthermore, drone warfare has profoundly impacted public opinion, leading to increased polarization and social division. While some sections of the population may view drones as a precise and effective tool for counterterrorism, others express deep moral and ethical objections to their use, citing concerns about civilian casualties and the potential for collateral damage. This stark division in public opinion has deepened existing societal fault lines and created fertile ground for political manipulation and exploitation. The debate surrounding drone warfare has become a powerful symbol of broader conflicts about the role of technology in warfare, the balance between national security and civil liberties, and the very nature of just war theory. The lack of a common understanding of the implications of drone warfare further complicates the task of finding common ground and fostering social cohesion.

The international implications of drone warfare are equally profound. The use of drones by certain nations has powered international tensions and prompted accusations of violating national sovereignty. The unilateral nature of drone strikes, often carried out in the territories of other countries without explicit consent or notification, has generated significant resentment and anger amongst affected governments. This has led to diplomatic disputes, strained international relations, and increased the risk of escalation and conflict. The perception that drone technology grants a military advantage to a select group of nations further exacerbates these geopolitical tensions, fostering an arms race and the potential for destabilization in international affairs.

Beyond the immediate diplomatic fallout, drone warfare raises significant human rights concerns. The lack of clear legal frameworks governing the use of drones in military operations, coupled with the challenges in verifying the identities of targets, increases the risk of civilian casualties and violates international humanitarian law. The potential for drone strikes to result in indiscriminate killings and the maiming of non-combatants has sparked widespread condemnation from human rights organizations. The psychological impact on civilian populations exposed to drone strikes is also significant, leading to widespread trauma, anxiety, and feelings of insecurity. These long-term consequences cannot be ignored when assessing the human cost of drone warfare.

The implications of drone warfare extend to the very core of democratic values. The increasing automation of lethal force, inherent in the use of drones, raises significant concerns about accountability and the erosion of human judgment in decision-making. The delegation of life-or-death decisions to algorithms raises profound ethical questions about the moral responsibility of states and the potential for the dehumanization of warfare. The loss of human agency in the targeting process undermines principles of due process and fair trial, blurring the line between lawful military operations and extrajudicial killings.

The case studies of several countries illuminate the far-reaching impacts of drone warfare. Yemen, for example, has experienced a devastating humanitarian crisis significantly exacerbated by the widespread use of drones in the ongoing conflict. The relentless drone strikes have resulted in numerous civilian casualties, causing immense human suffering and hampering the already fragile peace processes. Similarly, Pakistan, a country frequently targeted by drone strikes, has experienced significant societal and political upheaval as a result of these operations. The lack of transparency and accountability surrounding these strikes has eroded public trust in the government and fuelled anti-American sentiment. These examples highlight the profound and lasting consequences of drone warfare on civilian populations and their impact on political stability. Afghanistan, too, bears the scars of years of drone strikes, adding to the country's protracted instability and internal conflict.

As drone use expanded, so did critical scrutiny by investigative journalists and human rights organizations. The realities of civilian casualties, the lack of transparency surrounding targeting procedures, and the psychological impacts on affected populations became increasingly prominent in media narratives. The release of leaked documents, investigations into specific drone strikes, and eyewitness accounts painted a starkly different picture, revealing the human cost of drone warfare. The shift in media framing led to a noticeable decline in public support for the technology in several countries. Images of grieving families, destroyed homes, and scenes of trauma effectively countered the earlier narrative of surgical precision, humanizing the victims and forcing a broader reflection on the ethical implications of drone strikes.

The influence of media narratives is particularly pronounced in shaping the policy debate. Governments and military institutions frequently utilize strategic communications to manage public opinion and maintain support for drone programs. The emphasis often remains on the strategic benefits – counterterrorism successes, reduced military casualties, and cost-effectiveness – while downplaying the ethical dilemmas and potential for unintended consequences. This approach frequently meets resistance, particularly from media outlets that adopt a more critical stance, leading to an ongoing public discourse marked by varying narratives and conflicting interpretations of the facts. This constant media push and pull contributes to the politicization of drone technology, making it a highly contested issue in many public forums.

The contrast between the initial positive framing and the subsequent critical portrayals illustrates the power of media narratives in shaping public perception. The shift from technological marvel to ethical concern is not a simple change in information, but a fundamental alteration in the narrative's framing, resulting in a significant shift in public sentiment. The media's role in exposing government secrecy and highlighting instances of civilian casualties has been instrumental in galvanizing public dissent and pushing for greater accountability.

Furthermore, the accessibility of information through social media has played a decisive role in shaping public perception. Citizen journalism and the rapid dissemination of videos and images of drone strikes have bypassed traditional media gatekeepers, exposing a global audience to raw footage and personal accounts that often contradict official narratives. This immediacy and unfiltered nature of social media narratives have significantly amplified public concern and created a powerful platform for activism and debate. However, it has also created fertile ground for misinformation and the proliferation of biased content, posing a challenge in maintaining a nuanced understanding of complex events.

Effective communication and public engagement strategies are crucial for navigating this complex landscape. Transparency and accountability are paramount. Governments must strive for greater openness about drone programs, providing detailed information about targeting procedures and acknowledging civilian casualties. This transparency, while challenging in terms of national security concerns, is crucial for rebuilding public trust and fostering a more informed and responsible public discourse. Moreover, independent oversight mechanisms are essential for ensuring that drone operations comply with international humanitarian law and ethical guidelines.

The media itself has an essential role to play in fostering informed public debate. Responsible journalism requires a commitment to accurate reporting, balanced perspectives, and a critical examination of official narratives. Investigative reporting can unearth crucial information, while nuanced analyses can help the public understand the complexities of drone warfare beyond simplistic narratives. The media's commitment to fact-checking and the avoidance of sensationalism is particularly important in mitigating the spread of misinformation. Educational initiatives and public awareness campaigns can further contribute to a more informed public discourse by providing context, facts, and alternative perspectives.

Psychological Impact

The psychological strain on drone operators is a growing area of research. Unlike traditional warfare where combatants face immediate physical and psychological risks, drone operators are often situated thousands of miles away from the conflict zone, operating in sterile environments. This physical distance, however, does not equate to emotional detachment. Operators are often subjected to prolonged exposure to graphic images and sounds – the sights and sounds of death and destruction – relayed through the drone's sensors. This constant exposure to violence, often repeated over extended periods, can lead to a range of psychological consequences, including post-traumatic stress disorder (PTSD), moral injury, and burnout.

PTSD, traditionally associated with direct combat experiences, manifests in drone operators through persistent intrusive thoughts, flashbacks, nightmares, and hyper vigilance. The graphic nature of the imagery and the knowledge of their direct involvement in causing death and destruction can trigger these symptoms, even in the absence of direct physical danger. Studies have shown elevated rates of PTSD among drone operators compared to their non-deployed counterparts, highlighting the significant psychological burden associated with this type of warfare.

Furthermore, moral injury, a distinct psychological condition resulting from the perception of having violated one's own moral compass, is a significant concern for drone operators. The inherent disconnect between the remote, controlled environment of operation and the real-world consequences of their actions create fertile ground for moral conflict. Operators may grapple with the perceived lack of control over collateral damage, the difficulty in verifying targets, and the emotional weight of decisions leading to the loss of innocent lives. This sense of moral transgression can manifest as guilt, shame, anger, and a profound sense of disillusionment, severely impacting their mental and emotional well-being. The impersonal nature of drone warfare, often described as "video game warfare," can ironically amplify this moral injury, as the lack of physical immediacy can create a psychological distance that prevents necessary emotional processing and resolution.

The psychological impact extends beyond the direct experience of killing. The repetitive nature of the task, coupled with the emotional burden of witnessing death and destruction remotely, can contribute to burnout. The constant pressure to maintain situational awareness, the long working hours, and the emotional exhaustion from processing traumatic events can lead to decreased job satisfaction, cynicism, and a significant decline in mental health. This is further exacerbated by the lack of readily available support systems tailored to address the unique challenges faced by drone operators. The stigma associated with mental health issues within military culture often prevents individuals from seeking help, further compounding the problem. This silent suffering can have long-lasting consequences, affecting not just the individual operator but also their families and communities.

The psychological impact of drone warfare extends far beyond the operators. Civilians living in areas targeted by drone strikes experience a unique and devastating form of psychological trauma. The constant threat of death hanging over their heads, the unpredictable nature of strikes, and the resulting physical and emotional devastation create a pervasive climate of fear and anxiety. The psychological effects on civilian populations are often far-reaching and long-lasting, impacting mental health, social cohesion, and economic development. Witnessing a drone strike, or even hearing the characteristic drone sound, can trigger intense fear and panic, leading to symptoms consistent with PTSD and other anxiety disorders. The destruction of homes and infrastructure, the loss of loved ones, and the widespread displacement contribute to a sense of instability and insecurity, significantly impacting psychological well-being. Furthermore, the lingering fear and uncertainty, the lack of accountability, and the perceived powerlessness against this technologically superior adversary can generate deep-seated psychological distress that undermines community resilience and societal stability.

Children are particularly vulnerable to the psychological effects of drone warfare. The experience of witnessing a drone strike, or even living under the constant threat of one, can be profoundly traumatic for children. The fear, anxiety, and feelings of insecurity can severely impact their emotional development, leading to behavioural problems, sleep disturbances, and other mental health challenges. The loss of family members and the disruption of their normal lives contribute to a sense of grief and loss that can have lasting consequences. Moreover, the lack of access to psychological support services in many conflict zones means that these children often go without the necessary help to cope with their trauma.

The long-term psychological consequences of drone warfare on civilian populations extend beyond individual trauma. The erosion of trust in authority, the breakdown of social cohesion, and the widespread sense of insecurity can have far-reaching societal consequences. The fear and anxiety associated with drone strikes can hinder economic development and social progress. People may be reluctant to engage in economic activities, leading to decreased productivity and economic stagnation. The disruption of social structures, the displacement of families, and the breakdown of community support systems can exacerbate social inequalities and increase the risk of conflict.

Addressing the psychological impact of drone warfare requires a multi-pronged approach. For drone operators, improved mental health support systems, including access to specialized counselling and peer support groups, are crucial. The development of training programs that focus on ethical considerations, moral decision-making, and stress management can mitigate the risk of PTSD and moral injury. Transparency and accountability in targeting procedures are essential to reducing the psychological burden on operators, as well as fostering a sense of responsibility and justice. For civilian populations, access to mental health services, trauma-informed care, and psychosocial support are critical for addressing the psychological wounds inflicted by drone strikes. International humanitarian law must be rigorously observed, and efforts must be made to minimize civilian casualties and mitigate the psychological harm caused by drone warfare. This involves a commitment to accurate targeting, the thorough investigation of civilian casualties, and the provision of reparations to affected communities.

Furthermore, fostering a culture of open discussion and dialogue about the psychological consequences of drone warfare is essential. This involves encouraging operators, policymakers, and the public to acknowledge and address the ethical and psychological implications of this technology. Research into the long-term effects of drone warfare on both operators and civilian populations is needed to inform policy decisions and guide the development of effective mitigation strategies. International collaboration and cooperation are essential to establish common standards and guidelines for the ethical use of drone technology, minimizing the psychological harm inflicted on both combatants and non-combatants. The challenge lies in achieving a delicate balance between harnessing the technological capabilities of drones for security purposes and mitigating their potential for causing widespread psychological trauma. Only through a holistic and ethically informed approach can we hope to minimize the detrimental psychological consequences of this increasingly prevalent form of warfare.

The devastating immediate effects of drone warfare are readily apparent: the physical destruction of infrastructure, the loss of life, and the immediate trauma inflicted upon survivors. However, the long-term socioeconomic consequences are often less visible, yet equally profound and far-reaching. These consequences ripple through affected communities, hindering their ability to recover, rebuild, and ultimately, thrive. The disruption of economic activity, the erosion of social capital, and the lasting psychological scars all contribute to a complex web of interconnected challenges that impede sustainable peace and development.

One of the most significant long-term socioeconomic impacts is the disruption of economic activity. Drone strikes, even those targeting specific individuals or groups, often inflict collateral damage on civilian infrastructure. Markets, schools, hospitals, and agricultural lands can be destroyed or rendered unusable, leading to a sharp decline in economic productivity. The destruction of livelihoods is especially devastating for vulnerable populations who rely on small businesses or agriculture for their income. The resulting unemployment and poverty can fuel social unrest and instability, exacerbating existing inequalities and hindering long-term development. The destruction of essential infrastructure also disrupts supply chains, hindering access to essential goods and services, and creating further economic hardship. Reconstruction efforts, even when successful, often take years, and the economic scars can persist for decades, hindering the growth potential of affected regions.

Beyond the immediate physical destruction, drone warfare significantly undermines social capital. The constant threat of drone strikes creates a climate of fear and distrust, eroding the sense of security and community cohesion that is essential for social and economic progress. Families are displaced, social networks are fractured, and traditional support systems are weakened. This erosion of trust extends to governmental institutions as well. When civilians perceive a lack of accountability for drone strikes or a failure to protect them from harm, it can lead to a decline in public confidence and cooperation with authorities. This diminished trust makes it harder for governments to implement development programs and provide essential services, further hindering recovery and reconstruction efforts. The social fragmentation and loss of community support networks create an environment ripe for further conflict and instability.

The long-term psychological impact of drone warfare on civilian populations presents a further formidable challenge to socioeconomic recovery. The constant fear of death, the unpredictable nature of drone strikes, and the witnessing of violence all contribute to widespread trauma and mental health issues. The psychological consequences, including PTSD, anxiety disorders, and depression, affect individuals' ability to work, engage in social life, and contribute to the rebuilding of their communities. The intergenerational transmission of trauma is also a significant concern. Children who witness drone strikes or grow up in a climate of fear can experience lasting psychological damage, impacting their development, education, and future prospects. The resulting mental health crisis places significant strain on already limited healthcare systems and necessitates long-term investment in mental health services to address this pervasive issue.

Rebuilding war-torn regions after years of drone warfare presents a unique set of challenges. The physical reconstruction is a discouraging task, requiring significant resources and expertise. However, the process of rebuilding must go beyond the restoration of physical infrastructure. It must also address the deep-seated social, economic, and psychological wounds inflicted by the conflict. This requires a holistic approach that incorporates community-based development programs, targeted interventions to address mental health issues, and efforts to foster reconciliation and social cohesion. The need for sustainable economic development strategies that provide alternative livelihoods for displaced populations and create job opportunities is paramount.

Case studies from communities recovering from drone conflicts illustrate the complexities and challenges of this long-term rebuilding process. For example, in regions where agricultural land has been destroyed, community-based initiatives focusing on sustainable agricultural practices and providing access to microfinance have been implemented. These initiatives not only help to restore livelihoods but also promote community resilience and self-reliance. In areas where schools and hospitals have been damaged or destroyed, the rebuilding of infrastructure must be accompanied by programs that address the educational and healthcare needs of the population. The focus is often on training and education programs tailored to equipping individuals with the skills necessary to participate in the workforce. This may include vocational training programs, entrepreneurship training, and capacity-building initiatives to support the development of local businesses.

Addressing the long-term socioeconomic impacts requires a shift in the global discourse surrounding drone warfare. Open dialogue and critical discussion surrounding the ethical considerations and long-term effects of drone technology are crucial for informing future policy decisions and mitigating potential negative consequences. International cooperation and the establishment of clear guidelines governing the use of drones in armed conflict are necessary to minimize the risk of civilian casualties and mitigate the socioeconomic damage caused by drone strikes. This necessitates collaboration between governmental agencies, international organizations, and non-governmental organizations to establish mechanisms for monitoring, investigation, and accountability in cases of alleged civilian casualties. The development of international legal frameworks and mechanisms for redress and reparations to victims of drone strikes is crucial to fostering justice and accountability.

Ultimately, the long-term socioeconomic consequences of drone warfare are a complex and multifaceted challenge that demands a holistic and sustained response. Addressing these challenges requires not only significant financial investment but also a commitment to ethical considerations, sustainable development, and the well-being of affected communities. The goal should not only be to rebuild war-torn regions but also to create an environment where communities can heal, rebuild, and thrive in the long term. Only through such a comprehensive approach can we hope to mitigate the negative effects of drone warfare and promote sustainable peace and development in the affected regions. The path forward demands a global commitment to ethical use, transparency, and accountability in drone technology deployments, alongside comprehensive rehabilitation and development programs for affected communities. Only then can the long-term socioeconomic scars of drone warfare begin to heal.

The Future of Military Strategy and International Security

One of the most immediate concerns surrounding AWS is the potential for escalation. The speed and scale at which autonomous weapons can operate drastically reduce the time available for human decision-making. A rapid exchange of fire between opposing AWS, each reacting independently to perceived threats, could rapidly spiral into a full-scale conflict, leaving little room for diplomatic intervention or de-escalation efforts. The inherent unpredictability of complex autonomous systems, compounded by potential software glitches or adversarial manipulation, further amplifies this risk. A seemingly minor malfunction or a misinterpretation of data could trigger a catastrophic chain of events, potentially exceeding the thresholds for escalation previously defined by human actors. This inherent unpredictability poses a significant challenge to traditional notions of military deterrence, where the clear and controllable threat of retaliation plays a crucial role in conflict prevention.

Furthermore, the increasing autonomy of weapons systems raises serious concerns about accountability and the risk of unintended consequences. When a human operator is directly involved in the decision to engage a target, it is, in principle, possible to assign responsibility for any resulting civilian casualties or collateral damage. This becomes significantly more complicated with autonomous systems. Establishing accountability when a machine makes a life-or-death decision without direct human intervention presents formidable legal and ethical challenges. Determining culpability for malfunctions, software errors, or unintended targeting outcomes becomes a complex legal and technological puzzle. This lack of clear accountability could undermine international law and erode trust between nations, potentially leading to a proliferation of these weapons and an increased risk of their use in conflict.

The ethical implications of delegating life-or-death decisions to machines are equally profound. Existing international humanitarian law emphasizes the importance of human judgment and proportionality in the use of force. The inherent biases present in algorithms used to program AWS, coupled with the potential for unexpected behaviour in the face of complex situations, could undermine these principles. An AWS programmed to prioritize minimizing civilian casualties may still cause unintended deaths due to limitations in its sensor capabilities or its inability to fully understand the nuances of a battlefield situation. Conversely, an AWS designed with a less-discriminatory approach could cause significant civilian harm. The potential for such systems to disregard human values and international norms raises significant ethical concerns that must be addressed before widespread adoption.

The development and deployment of AWS are also likely to have a significant impact on military doctrines and strategies. The speed and precision of autonomous weapons could revolutionize warfare, enabling rapid, decisive strikes with potentially minimal human involvement. This could lead to a shift in military thinking, favouring smaller, highly-trained Special Forces supported by autonomous weapons over large-scale conventional armies. Existing doctrines designed for human-centric warfare would need significant revision to incorporate the capabilities and limitations of AWS. Training regimes would require substantial modifications, shifting emphasis towards understanding and managing autonomous systems rather than traditional combat skills. Furthermore, the reliance on autonomous systems could introduce vulnerabilities. A successful cyber attack targeting these systems could have devastating consequences, disrupting military operations and potentially causing significant loss of life.

The potential impact of AWS on international security extends beyond the battlefield. The development and proliferation of these systems could lead to a new arms race, with nations competing to develop increasingly sophisticated and autonomous weapons. This arms race could destabilize the international order, increasing the risk of accidental or intentional conflict. The potential for these systems to fall into the hands of non-state actors, such as terrorist organizations, further exacerbates the risks. An AWS in the hands of a non-state actor poses a far greater threat than conventional weapons, given its ability to operate autonomously and inflict significant damage without the need for extensive training or logistical support. The lack of effective international mechanisms for controlling and regulating the development and use of AWS makes the prospect of such an eventuality deeply concerning.

Artificial Intelligence (AI) Integration

Current UCAVs rely heavily on human operators for decision-making, even in semi-autonomous modes. Future generations, however, will likely feature increasingly sophisticated AI systems capable of autonomous target recognition, engagement prioritization, and even strategic decision-making within pre-defined parameters. This increased autonomy will not only enhance the speed and efficiency of operations but will also enable UCAVs to react more swiftly to dynamic battlefield situations. The potential benefits are significant: quicker response times to emerging threats, improved situational awareness through enhanced data processing, and reduced reliance on human operators in high-risk environments. This enhanced autonomy, however, raises ethical and legal concerns that must be addressed proactively. The development of robust ethical guidelines and stringent oversight mechanisms is paramount to ensure responsible and accountable use of AI-enabled UCAVs.

Autonomous operation is intrinsically linked to advancements in sensor technology. Current UCAVs rely on a combination of sensors, including electro-optical/infrared (EO/IR) cameras, synthetic aperture radar (SAR), and electronic intelligence (ELINT) systems. Future systems will likely integrate more advanced sensors with improved resolution, range, and multi-spectral capabilities. Hyper spectral imaging, for instance, will offer far more detailed analysis of targets and their surroundings, improving target discrimination and reducing the risk of collateral damage. The integration of advanced data fusion algorithms will further enhance situational awareness, enabling UCAVs to more effectively process and interpret data from multiple sources. This will allow for a more comprehensive and accurate understanding of the battlefield, enabling more informed decision-making. This advanced sensor suite will also allow for the development of specialized UCAVs with capabilities tailored to specific missions, such as those focused on chemical, biological, radiological, and nuclear (CBRN) threat detection or those designed for maritime surveillance and anti-piracy operations.

Swarm technology represents another significant area of development. The concept involves coordinating large numbers of relatively inexpensive and relatively simple UCAVs to overwhelm enemy defenses through sheer numbers and coordinated action. These swarms could operate independently or in conjunction with more sophisticated, individually controlled UCAVs. Such a strategy presents a new paradigm in warfare, potentially rendering traditional point-defense systems less effective. The coordinated actions of a swarm can overwhelm an adversary's ability to engage each drone individually, saturating defenses and achieving mission objectives with a high degree of success. The operational implications of swarm technology are profound. It could fundamentally alter the balance of power, allowing smaller, less technologically advanced forces to pose a significant threat to more technologically advanced adversaries. Furthermore, swarms could be employed for a wide variety of tasks, including reconnaissance, electronic warfare, precision strikes, and even disabling critical infrastructure.

The future of UCAV design will also see significant changes. Stealth technology, already employed in some advanced manned aircraft, will become increasingly prevalent in UCAV design. Reduced radar cross-section (RCS) will make UCAVs harder to detect and track, improving their survivability and effectiveness. Furthermore, the integration of advanced materials, such as carbon fibber composites, will result in lighter, more durable, and more fuel-efficient airframes. This will translate to longer flight times, greater payload capacity, and increased operational range. The potential for hypersonic UCAVs also represents a significant area of research and development. Hypersonic speeds would allow for incredibly rapid deployment and engagements, rendering many current air defence systems ineffective. However, the technological challenges associated with hypersonic flight are considerable, requiring breakthroughs in propulsion, materials science, and thermal management. The development of hypersonic UCAVs will likely take considerable time and substantial investment but would provide a significant tactical and strategic advantage if successful.

However, the development of these advanced UCAV capabilities also introduces new challenges. The potential for AI-enabled systems to malfunction or to be exploited by adversaries presents a significant risk. Robust cyber security measures are critical to prevent unauthorized access and control, ensuring the safe and responsible operation of these advanced systems. Moreover, the ethical implications of autonomous weapons systems require careful consideration and international cooperation. Establishing clear legal frameworks and international norms governing the development and use of autonomous weapons is essential to prevent unintended escalation and ensure the responsible use of this technology.

Looking further ahead, the convergence of UCAV technology with other unmanned systems, such as unmanned ground vehicles (UGVs) and unmanned underwater vehicles (UUVs), will create new operational possibilities. Imagine coordinated operations involving UCAVs providing aerial surveillance and reconnaissance, while UGVs provide ground-based support and UUVs conduct underwater reconnaissance. This synergistic approach will provide a much more comprehensive and effective operational picture, significantly enhancing situational awareness and decision-making capabilities. This multi-domain operational approach could revolutionize the conduct of modern warfare.

The advancements in UCAV technology and operational concepts discussed above will inevitably have significant implications for military doctrine and strategy. Military forces will need to adapt their operational procedures, training programs, and organizational structures to effectively integrate these advanced technologies. The development of new tactics, techniques, and procedures (TTPs) will be vital for maximizing the effectiveness of UCAVs while mitigating their associated risks. Furthermore, the integration of UCAVs into larger military networks and command and control systems will be critical for effective coordination and interoperability.

The integration of artificial intelligence (AI) represents a quantum leap in drone technology, transforming them from remotely piloted vehicles into increasingly autonomous systems capable of complex tasks with minimal human intervention. This integration isn't simply about automating existing functions; it's about fundamentally altering the capabilities and operational potential of drones. AI empowers drones to adapt to dynamic environments, make real-time decisions, and execute missions with a level of sophistication previously unimaginable.

One of the most significant impacts of AI is in the realm of autonomous flight control. Traditional drone control systems rely heavily on pre-programmed flight plans and operator input. However, AI-powered systems enable drones to navigate complex and unpredictable environments with far greater autonomy. Machine learning algorithms allow drones to learn from vast datasets of flight data, improving their ability to handle unexpected conditions like strong winds, obstacles, and GPS interference. These algorithms enable the drones to adjust their flight path in real-time, ensuring safe and efficient navigation even in challenging circumstances. For example, a drone equipped with an AI-powered obstacle avoidance system can autonomously navigate a cluttered urban environment, detecting and avoiding buildings, trees, and power lines without human intervention. This capability is crucial for applications such as search and rescue operations, infrastructure inspections, and delivery services in densely populated areas.

Beyond basic navigation, AI enables drones to perform more complex flight manoeuvres. AI algorithms can optimize flight paths for fuel efficiency, minimizing energy consumption and maximizing flight range. They can also coordinate the movements of multiple drones in a swarm, enabling complex formations and synchronized actions. This capability is especially critical for military applications, where coordinated drone swarms can overwhelm enemy defences or execute complex tactical manoeuvres. Imagine a swarm of AI-powered drones autonomously coordinating their movements to suppress enemy air defences before a manned aircraft strike, or working together to map a disaster zone with high precision and speed. The implications are transformative for military strategy and tactics.

Furthermore, AI is revolutionizing the way drones interact with their environment through advanced sensor fusion. Modern drones are equipped with a variety of sensors, including cameras, LiDAR, radar, and infrared sensors. AI algorithms can integrate data from multiple sensors to create a comprehensive understanding of the surrounding environment, allowing drones to make more informed decisions. For instance, a drone equipped with an AI-powered sensor fusion system can accurately identify and classify objects in a scene, even in low-visibility conditions. This capability is central for tasks such as target recognition, surveillance, and search and rescue operations. AI algorithms can differentiate between different types of vehicles, people, and objects, providing critical information to operators in real-time. This real-time analysis of sensory data empowers operators to make faster, more informed decisions, saving valuable time and potentially lives in critical situations.

The integration of AI also significantly enhances the decision-making capabilities of drones. AI algorithms can analyze vast amounts of data from multiple sources to identify patterns and make predictions, allowing drones to react more quickly and effectively to changing situations. For example, AI can analyze sensor data to predict the likelihood of an imminent threat, triggering an automated evasive manoeuvre or defensive countermeasure. This capability is crucial for military applications, where drones may be deployed in high-risk environments. But the implications extend beyond the military; autonomous decision-making is central for safe and efficient operation in civilian contexts such as autonomous delivery systems, where drones must navigate complex environments while avoiding collisions and making intelligent delivery decisions.

Machine learning, a subset of AI, plays a central role in enhancing several key aspects of drone operation. One significant application is in target recognition. Through training on large datasets of images and videos, machine learning algorithms can learn to identify specific targets with high accuracy. This capability has profound implications for military applications, enabling drones to autonomously identify and engage enemy targets with increased precision and reduced risk to human operators. Furthermore, the ability to identify friendly forces is equally critical, avoiding accidental fratricide. The continuous learning and improvement capabilities of machine learning algorithms allows for ever more precise and reliable target identification as the systems are exposed to more data.

Machine learning also improves obstacle avoidance capabilities. By analyzing sensor data and learning from past experiences, AI algorithms can develop sophisticated obstacle avoidance strategies that are more robust than traditional rule-based systems. This is particularly important for drones operating in complex and unpredictable environments. Consider the challenges of autonomous flight through dense forests or urban canyons, requiring a level of adaptability that surpasses traditional programming. Machine learning enables drones to learn to handle these complexities dynamically.

Moreover, machine learning is used to optimize path planning. AI algorithms can generate optimal flight paths, considering factors such as terrain, weather conditions, and obstacles. This optimization not only enhances efficiency but also improves the safety and reliability of drone operations. For example, in search and rescue missions, an AI-powered path planning system can quickly generate the most efficient route to the target location, minimizing the time it takes to reach those in need. In situations with time-sensitive emergencies, this optimization is paramount.

The advancements in AI are not without their challenges. One major concern is the potential for bias in AI algorithms. If the data used to train AI algorithms is biased, the resulting algorithms may also be biased, leading to unfair or discriminatory outcomes.

Addressing this issue requires careful consideration of data selection and algorithm design. Moreover, ensuring the robustness and security of AI systems is critical. AI algorithms can be vulnerable to adversarial attacks, where malicious actors attempt to manipulate the algorithms to produce incorrect results. Robustness and security measures are vital to mitigating this risk. Finally, the ethical implications of AI-powered autonomous drones must be carefully considered. The potential for autonomous lethal weapons raises significant ethical concerns that require careful consideration and international cooperation to establish ethical guidelines and regulations.

The integration of AI is transforming drone technology in profound ways. From autonomous flight control and advanced sensor fusion to sophisticated decision-making algorithms, AI is empowering drones to perform increasingly complex tasks with minimal human intervention. While challenges remain, the potential benefits of AI-powered drones are vast, spanning various sectors from military applications to civilian uses, promising a future where drones play an increasingly vital role in our lives. The continuing evolution of AI and its integration into drone technology will undoubtedly shape the future landscape of both military and civilian applications, necessitating ongoing dialogue regarding its ethical and societal implications. The ongoing research and development in this field promise further advancements, pushing the boundaries of what's possible with unmanned aerial systems.

The transformative power of AI in drone technology extends far beyond autonomous flight and decision-making; it fundamentally reshapes how drones perceive and interact with their environment. This hinges critically on the sophisticated integration of diverse sensor technologies and the advanced algorithms that fuse their data streams into a cohesive and actionable understanding of the surrounding world. Modern unmanned aerial vehicles (UAVs) are no longer reliant on single-sensor solutions; instead, they employ a suite of sensors, each offering a unique perspective and data modality, to create a comprehensive and robust situational awareness picture.

One of the cornerstone sensor technologies is LiDAR (Light Detection and Ranging). LiDAR uses pulsed laser light to measure distances to objects, creating a three-dimensional point cloud representation of the environment. This technology excels at mapping terrain, detecting obstacles, and generating high-resolution 3D models of structures. The precision of LiDAR allows drones to navigate complex environments with exceptional accuracy, particularly in GPS-denied areas or when operating in challenging weather conditions. For instance, in search and rescue operations within dense forests, LiDAR can penetrate foliage, providing a detailed map of the terrain below, aiding in locating survivors and navigating the difficult landscape. Moreover, LiDAR's ability to create accurate digital elevation models (DEMs) is invaluable for infrastructure inspections, enabling the detection of damage or anomalies in bridges, power lines, and other critical infrastructure. The high data density of LiDAR presents computational challenges, requiring robust processing power onboard the UAV or for post-mission analysis.

Radar, another vital sensor type, offers distinct advantages in challenging conditions. Unlike LiDAR, radar is not affected by light conditions and can penetrate fog, rain, and smoke. Using radio waves, radar provides information on the range, velocity, and direction of objects, irrespective of visibility. This is particularly critical for military applications, where drones may be deployed at night or in adverse weather. Radar's ability to detect moving targets is vital for surveillance, target tracking, and threat avoidance. Several types of radar are used in UAVs, including synthetic aperture radar (SAR), which can create high-resolution images of the ground, even in the absence of direct sunlight. Ground-penetrating radar (GPR) is also employed for subterranean target detection, offering capabilities that extend beyond surface-level observation. The data from radar is often complementary to data from other sensors, enhancing the overall situational awareness of the drone.

Electro-optical (EO) and infrared (IR) sensors form the backbone of many UAV-based surveillance and reconnaissance missions. EO cameras provide high-resolution visual imagery, similar to a conventional camera, while IR sensors detect heat signatures. The combined use of EO and IR sensors significantly enhances target identification and classification capabilities. For example, IR imagery can reveal hidden heat sources, such as concealed vehicles or individuals, which might be obscured in visible light imagery. This fusion of data streams allows for improved object recognition and contextual understanding, distinguishing between potential threats and benign objects. Furthermore, EO/IR systems are frequently integrated with image processing algorithms that enhance image quality and automatically identify potential targets of interest, reducing the cognitive load on human operators.

Beyond these core sensor types, UAVs increasingly incorporate other sensors to enhance their capabilities. These include magnetometers for detecting magnetic anomalies, acoustic sensors for listening to sounds, and even chemical sensors for detecting airborne pollutants or hazardous substances. The versatility of sensor payloads allows drones to be adapted to a wide range of applications, from environmental monitoring to precision agriculture to law enforcement. The true power of these diverse sensor technologies, however, lies not in their individual capabilities but in their synergistic integration through data fusion. Data fusion algorithms process information from multiple sensors, resolving inconsistencies, and creating a more robust and comprehensive understanding of the environment than any single sensor could provide on its own.

This process often involves complex mathematical techniques, such as Kalman filtering, which integrates sensor data over time to reduce noise and improve accuracy. Probabilistic methods are also employed to assess the reliability of different sensors and weight their contributions accordingly, enabling the system to manage uncertainties inherent in sensor data.

One of the significant challenges in sensor fusion is managing the vast amounts of data generated by multiple sensors. The real-time processing of this information requires considerable computational power. The development of lightweight, high-performance processors is essential for making sensor fusion feasible in small, battery-powered UAVs. Furthermore, the algorithms themselves must be optimized for efficiency, minimizing latency and maximizing data throughput. Algorithms must account for the inherent biases and limitations of individual sensors and compensate for these imperfections in the fusion process.

Furthermore, the robustness and reliability of sensor fusion algorithms are vital. The system must be resilient to sensor failures or malfunctions, ensuring that the overall situational awareness is not compromised if one sensor fails. Fault tolerance and graceful degradation are essential considerations in the design of data fusion systems for autonomous UAVs. The potential consequences of sensor fusion failure are significant, ranging from navigational errors to mission failures, and ultimately, safety hazards.

The continued advancement of sensor technologies and data fusion algorithms will significantly impact the capabilities of autonomous drones. Miniaturization of sensor components, along with improved computational power, will enable the development of smaller, lighter, and more cost-effective UAVs. The integration of novel sensor modalities, such as hyperspectral imaging and synthetic aperture sonar, will further expand the range of applications for these unmanned systems. Ongoing research in machine learning and artificial intelligence is leading to the development of more sophisticated data fusion algorithms capable of handling even greater volumes of data and extracting higher-level knowledge from raw sensory information. This leads to enhanced object recognition, threat detection, and decision-making capabilities, pushing the boundaries of what's possible with autonomous drones.

Future of Autonomous Drone Technology

The future of autonomous drone technology is intrinsically linked to the ongoing advancements in sensor technologies and data fusion. As sensors become smaller, more powerful, and more diverse, and as data fusion algorithms become more sophisticated and robust, autonomous UAVs will play an increasingly prominent role in both military and civilian contexts. The integration of these capabilities paves the way for a future where drones operate with unprecedented levels of autonomy and situational awareness, undertaking tasks previously considered beyond the realm of possibility. This progress, however, also necessitates thoughtful consideration of the ethical and societal implications of this technological advancement, ensuring responsible development and deployment of these increasingly capable systems.

The ability of a drone to effectively utilize this sensory information and execute its mission relies heavily on its communication capabilities. The complex world of communication protocols and network architectures used in UAV operations, focusing on the challenges of maintaining reliable and secure communication links across diverse and often unpredictable environments. The effectiveness of any autonomous drone system hinges on the seamless and timely exchange of data. This involves not just the transmission of sensor data, but also commands from the ground control station (GCS), navigational updates, and situational awareness information shared between multiple drones in a swarm. The communication infrastructure must be robust enough to handle these data streams, even under challenging conditions such as signal interference, geographical limitations, and adversarial jamming attempts.

Line-of-sight (LOS) communication, the simplest form, uses direct radio links between the UAV and the GCS. This approach is highly effective when the UAV remains within the direct line of sight of the ground station, offering high bandwidth and low latency. However, LOS communication's range is limited by the curvature of the earth and obstacles like buildings, terrain, and even weather conditions like heavy rain or fog. Consequently, LOS is unsuitable for long-range or extended missions where the UAV may operate beyond the direct visual range of the operator. Furthermore, while LOS offers high bandwidth and low latency, it's inherently vulnerable to interception and jamming attacks, raising significant security concerns, particularly in contested environments. In military applications, for instance, an adversary could readily disrupt communication, leading to a loss of control or compromised mission success.

To overcome the limitations of LOS communication, various beyond-line-of-sight (BLOS) techniques have been developed. These techniques are crucial for enabling extended-range operations and autonomous missions that venture far from the GCS. One prominent method involves employing cellular networks (3G, 4G, 5G, and beyond) to establish communication links. Cellular networks provide widespread coverage, making them attractive for various applications, including search and rescue, infrastructure inspection, and delivery services. However, the bandwidth offered by cellular networks might be insufficient to support the high-data-rate requirements of advanced sensor systems and autonomous decision-making algorithms. Bandwidth limitations are especially noticeable when the drone is required to transmit high-resolution imagery or LIDAR point clouds.

Satellite communication offers a solution for truly long-range BLOS operations, providing global coverage and a degree of resilience against terrestrial interference. Satellite constellations dedicated to UAV communication are emerging, specifically designed to provide high-bandwidth, low-latency connections to drones operating in remote areas or beyond the reach of terrestrial networks. This opens up diverse possibilities for military and civilian uses, facilitating global surveillance, environmental monitoring, and disaster relief efforts in previously inaccessible locations. However, the cost of satellite communication can be substantial, and the latency can be higher compared to terrestrial communication systems, impacting real-time control and data transmission. Furthermore, reliance on satellites introduces vulnerabilities associated with satellite availability, signal blockage, and potential adversarial interference with satellite links.

Mesh networking is an increasingly important communication paradigm for multi-UAV operations, particularly for swarms or collaborative missions. In a mesh network, drones communicate not only with the GCS but also with each other, creating a distributed and resilient communication infrastructure. If one communication link is disrupted, the data can be routed through other drones in the network, ensuring communication continuity. This inherent redundancy greatly improves the robustness of the system and its resistance to jamming and interference. Mesh networking is particularly valuable in environments with limited terrestrial infrastructure or where signal propagation is unreliable, such as dense urban areas or mountainous terrain. However, the management and coordination of a mesh network can be complex, requiring sophisticated protocols and algorithms to efficiently manage data routing and network topology. The security of mesh networks also requires careful consideration, as vulnerabilities in any node can compromise the entire network.

Improving the robustness and security of drone communication is vital. Various techniques are employed to achieve this. Error correction codes are routinely used to detect and correct errors introduced during data transmission over noisy channels. Forward error correction (FEC) techniques add redundancy to the data stream, allowing the receiver to reconstruct the original message even if some parts are lost or corrupted during transmission. Data encryption and authentication protocols are essential for ensuring the confidentiality and integrity of transmitted data, particularly in sensitive applications such as military reconnaissance or critical infrastructure monitoring. Techniques like AES (Advanced Encryption Standard) and digital signatures are employed to protect communication against unauthorized access and tampering. Furthermore, frequency hopping and spread spectrum techniques are used to mitigate the impact of jamming attacks by making it difficult for an adversary to effectively disrupt the communication link.

The choice of communication protocol and network architecture for a given drone application depends on several factors, including the mission's range, operational environment, data rate requirements, security considerations, and the number of drones involved. A small UAV performing local surveillance might rely on LOS communication, while a large swarm of drones engaged in a complex coordinated operation would require a robust and flexible mesh network augmented by satellite or cellular links. The development of advanced, adaptive communication systems that can dynamically switch between different communication modalities and adjust their parameters based on the prevailing environmental conditions is a critical area of ongoing research. Future advancements in communication technology, including improved satellite constellations, higher-bandwidth cellular networks, and more efficient mesh networking protocols, will further enhance the capabilities and reliability of autonomous drone systems. The ongoing integration of artificial intelligence into communication systems also promises to improve resilience, efficiency, and security by enabling autonomous network management, adaptive routing, and intelligent interference mitigation. As drones continue to take on increasingly complex roles, the development of secure, robust, and adaptive communication architectures remains a cornerstone of this rapidly evolving technology.

The rapid technological progress has outpaced the development of a robust ethical and legal framework to govern their deployment and use. This creates a significant void, raising profound questions about accountability, potential misuse, and the need for international cooperation in establishing clear guidelines and regulations.

The development of "explainable AI" (XAI) is another essential aspect in addressing the ethical and legal challenges associated with autonomous drones. XAI systems are designed to provide insight into the decision-making process of AI algorithms, allowing users and investigators to understand the reasoning behind the system's actions. This enhanced transparency can help to build trust and improve accountability, making it easier to identify potential biases or errors in the system's algorithms. The integration of XAI into autonomous drones can significantly improve the ability to assess the legality and ethical implications of their actions.

Furthermore, the development of effective kill switches or override mechanisms is vital. While some argue that a fully autonomous system should be able to operate independently without human intervention, the potential for unforeseen circumstances and unintended consequences necessitate the inclusion of safeguards that allow human operators to intervene if necessary. Such mechanisms must be reliable, easily accessible, and capable of effectively halting the drone's operation, even in challenging operational environments. The design and implementation of these kill switches should incorporate redundancy and robust security protocols to prevent unauthorized access or manipulation.

The debate surrounding autonomous weapons systems, often referred to as lethal autonomous weapons (LAWs), is particularly heated. The prospect of machines making life-or-death decisions without meaningful human intervention raises serious ethical and legal concerns. International discussions are underway to establish norms and possibly a treaty banning the development and deployment of LAWS. The central argument against LAWS focuses on their potential to dehumanize warfare, reduce accountability, and increase the risk of unintended escalation. Proponents, however, argue that LAWS could increase precision and reduce civilian casualties compared to human-controlled systems, though this remains a contentious point.

Hypersonic and High-Speed Drones

The development of high-speed and hypersonic drones presents significant challenges in aerodynamics and propulsion. Unlike subsonic drones, which operate within the relatively predictable realm of airflow, high-speed drones must contend with the complexities of supersonic and hypersonic flight regimes. These regimes introduce phenomena such as shock waves, boundary layer separation, and extreme heating, all of which demand innovative engineering solutions.

One of the primary aerodynamic considerations is the design of the airframe. At supersonic speeds, the airflow around the drone becomes highly compressed, leading to the formation of shock waves. These shock waves can cause significant drag and instability, impacting the drone's manoeuvrability and efficiency. To mitigate these effects, high-speed drone designs often incorporate features such as slender bodies, swept wings, and sharp leading edges. These features help to reduce drag and control shock wave formation, ensuring stability and efficient flight. The shape of the drone becomes critical; a blunt body, for example, will generate a much stronger bow shock wave than a streamlined design. Computational fluid dynamics (CFD) simulations play a critical role in optimizing these designs, allowing engineers to test and refine various airframe configurations in a virtual environment before physical prototyping, significantly reducing development costs and time.

Beyond the airframe, the control surfaces are also critical for high-speed flight. At supersonic speeds, conventional control surfaces may become ineffective due to the high aerodynamic forces involved. Therefore, high-speed drones often incorporate advanced control systems, such as thrust vectoring or aerodynamic control surfaces specially designed for supersonic or hypersonic flow conditions. These control systems use multiple actuators, providing fine-tuned control over attitude and trajectory. Sophisticated flight control algorithms are necessary to manage the increased complexity and instability inherent in high-speed flight and ensure stable and precise control. The control software often incorporates adaptive control techniques, allowing the drone to adjust its flight parameters in real-time based on sensed environmental conditions and prevailing aerodynamic forces.

The extreme heating experienced at hypersonic speeds presents another significant challenge. As a drone travels at hypersonic speeds, the friction between the airframe and the atmosphere generates immense heat. This heat can cause significant structural damage, potentially leading to catastrophic failure. To address this, high-speed and hypersonic drones often incorporate advanced thermal protection systems (TPS). These systems use specialized materials and coatings to dissipate or absorb heat, protecting the drone's internal components from extreme temperatures. Materials such as ablative materials, which vaporize to absorb heat, and ceramic matrix composites (CMCs), which offer high strength and thermal resistance, are frequently used in TPS designs. These materials are often tailored to specific hypersonic flight conditions and can include internal cooling systems that utilize advanced fluids or gas coolants to effectively dissipate heat.

Propulsion systems for high-speed drones are equally critical. Traditional propeller-based systems are inadequate for supersonic and hypersonic flight. Instead, high-speed drones typically utilize jet engines or rockets, depending on the desired speed and mission profile. Ramjets, scramjets, and rocket engines each possess unique characteristics that make them suitable for different applications. Ramjets, for instance, are efficient at supersonic speeds but become less efficient at subsonic speeds, requiring a different propulsion system for takeoff and lower-speed manoeuvres. Scramjets offer the potential for even higher speeds, operating efficiently in the hypersonic regime.

Ramjet engines rely on the forward motion of the drone to compress the incoming air, mixing it with fuel and igniting it for thrust. They are relatively simple in design, but their reliance on high-speed airflow limits their applicability to low altitudes. Scramjets, on the other hand, operate by compressing incoming air at hypersonic speeds without the need for a mechanical compressor, offering the possibility of even greater speeds and higher altitudes compared to ramjets. The design and operational challenges of scramjets are significantly greater compared to ramjets, however, requiring advanced materials and precise control systems to manage the extreme temperatures and pressures involved. Rocket engines, meanwhile, offer the highest thrust-to-weight ratio, making them ideal for short bursts of high acceleration or high-altitude flight, but are less fuel-efficient for sustained flight than ramjets or scramjets.

The selection of propulsion system often involves trade-offs between speed, range, fuel efficiency, and operational complexity. A hypersonic drone designed for a short-range, high-speed strike mission may utilize a rocket engine for its high thrust, while a drone designed for reconnaissance or persistent surveillance might prioritize fuel efficiency and opt for a ramjet or scramjet engine. The integrated design, optimizing the airframe and propulsion system, is a complex optimization challenge, taking into account the interplay between aerodynamics, thermal management, and propulsion efficiency.

Research and development efforts in high-speed drone propulsion are actively pursuing advancements in several key areas. These include improved combustion efficiency in ramjets and scramjets, leading to longer flight durations and increased range; development of more efficient and lightweight fuel sources; and advancements in materials science, focusing on high-temperature materials capable of withstanding the extreme thermal stresses experienced at hypersonic speeds. The integration of advanced propulsion systems with sophisticated control algorithms further enhances the capabilities of these drones, enabling precise manoeuvrability at extremely high speeds.

One notable example of ongoing research is the development of hybrid propulsion systems that combine the advantages of different engine types. A drone might utilize a rocket booster for initial launch and acceleration, switching to a scramjet for sustained hypersonic flight. Such hybrid systems offer a greater degree of flexibility and adaptability, enabling a broader range of mission profiles.

Furthermore, the miniaturization of propulsion systems is also a significant area of focus. Smaller, more compact engines are vital for reducing the size and weight of high-speed drones, increasing their manoeuvrability and stealth capabilities. Research in micro-combustors and other miniaturization technologies promises to revolutionize the propulsion systems available for smaller high-speed drones. This miniaturization also opens up opportunities for the development of smaller, more manoeuvrable hypersonic drones with enhanced stealth capabilities.

The challenges inherent in designing and building high-speed and hypersonic drones are considerable. The demanding aerodynamic and propulsion requirements necessitate advanced engineering techniques, innovative materials, and sophisticated control systems. However, the potential benefits, including enhanced speed, range, and manoeuvrability, make the continued pursuit of this technology a high priority for both military and civilian applications. The advancements in computational fluid dynamics, materials science, and propulsion systems are crucial in overcoming these challenges and realizing the full potential of high-speed and hypersonic drones in the coming decades. The development of this technology also presents opportunities for spin-off technologies, advancing our understanding of high-speed flight and materials science, ultimately benefiting other sectors such as aerospace engineering and even high-speed transportation. The international race to dominate this field underscores its strategic importance and the ongoing investment in research and development worldwide.

The extreme temperatures generated during hypersonic flight pose a formidable challenge to the structural integrity of high-speed drones. Unlike subsonic flight, where thermal considerations are relatively less critical, hypersonic speeds induce intense frictional heating, potentially exceeding several thousand degrees Celsius at the leading edges and surfaces of the airframe. This necessitates the use of advanced materials and sophisticated thermal management systems to ensure the drone's structural integrity and the functionality of its onboard systems. The selection of materials is not a simple process but rather a complex optimization problem considering factors like weight, strength, thermal resistance, cost, and manufacturing feasibility.

One of the key areas of focus in materials science for hypersonic drones is the development of high-temperature structural materials. Traditional aerospace alloys, while suitable for subsonic and even supersonic speeds, often lack the necessary thermal resistance for hypersonic flight. This has driven intensive research into advanced materials like ceramic matrix composites (CMCs). CMCs are composite materials composed of ceramic fibbers embedded within a ceramic matrix. This structure combines the high-temperature strength and thermal shock resistance of ceramics with the toughness and damage tolerance of the matrix material. Silicon carbide (SiC) and carbon fibbers are commonly used reinforcements, offering exceptional strength and high-temperature capabilities. The matrix can be composed of various ceramics, each offering a different balance of properties. The choice of fibber and matrix is tailored to specific requirements dictated by the drone's mission profile, the design constraints of the airframe, and the expected thermal loads.

Another class of materials vital for hypersonic applications are ablative materials. These materials are designed to absorb heat through vaporization or decomposition, thus protecting the underlying structure. They are often used in areas of the airframe that experience the most intense heating, such as the leading edges of wings and control surfaces. Ablative materials can be polymers, composites, or even specially treated ceramics. Their effectiveness is determined by their heat absorption capacity, the rate at which they ablate, and their ability to shed away the ablated material without causing further damage or instability. The controlled ablation process ensures that the bulk temperature of the material remains below a critical threshold, preventing catastrophic failure. The design of ablation systems must carefully consider the mass loss during flight, as this affects the drone's aerodynamic properties and centre of gravity. Sophisticated modelling and simulation techniques are employed to predict ablation rates and optimize the design of ablative components.

Beyond the structural materials, the selection of materials for internal components is also critical. Onboard electronics, sensors, and actuators must be able to operate reliably under extreme thermal stresses. This often requires the use of specialized electronic components designed for high-temperature applications, such as high-temperature electronics packaging materials and radiation-hardened integrated circuits. The thermal management of these internal components requires equally advanced techniques. Passive cooling methods, such as heat sinks and thermal barriers, are often insufficient at hypersonic speeds, leading to the requirement for active cooling systems.

Active cooling systems for hypersonic drones employ various techniques to dissipate heat. One common approach is the use of liquid or gas coolants that circulate through passages within the structure, transferring heat away from critical components. This requires advanced fluid dynamics modelling and precise control of the coolant flow to ensure efficient heat transfer without compromising the overall structural integrity. The choice of coolant depends on its thermal properties, its compatibility with the materials used in the system, and its overall density, among other factors. The development of high-temperature coolants is a significant area of ongoing research, with materials such as molten salts and specialized fluids being investigated for their potential use in these extreme environments.

The development and implementation of effective thermal protection systems (TPS) are critical for the success of hypersonic drone programs. TPS are designed to minimize the heat transferred to the airframe and its internal components. The design of a TPS involves careful consideration of the material properties, the geometry of the thermal protection components, and the integration with the overall airframe design. The material selection must balance thermal resistance with weight and aerodynamic considerations. TPS are often layered, with each layer performing a specific function, such as radiation shielding, heat absorption, and heat dissipation.

Advanced computational techniques, specifically Computational Fluid Dynamics (CFD) simulations and Finite Element Analysis (FEA), play a vital role in designing and optimizing both the airframe and the thermal protection systems. CFD simulations predict the airflow and heat transfer around the drone, enabling engineers to assess the thermal loads on different parts of the airframe and optimize the TPS design for maximum effectiveness. FEA simulations then allow for the prediction of the structural response to these thermal loads, ensuring that the airframe can withstand the extreme temperatures and stresses without compromising its structural integrity. These simulations also consider the coupling effects between the airframe, the TPS, and the internal components, producing more accurate predictions of the overall system performance under hypersonic flight conditions.

The testing and qualification of materials and components for hypersonic drones require specialized facilities and experimental methods. Ground-based testing often involves the use of arc jets, plasma torches, and other high-heat-flux facilities that can simulate the intense thermal environment experienced during hypersonic flight. These tests are conducted on small-scale models or representative samples of materials and components to evaluate their thermal resistance, ablation characteristics, and overall durability. Flight testing, however, is essential to validate the performance of the TPS in a real-world environment. This involves equipping test drones with extensive instrumentation to measure temperatures, pressures, and other parameters during hypersonic flight. This data is then used to refine the materials selection and the TPS design, enabling continuous improvement of the technology.

The ongoing research and development in materials science and thermal management for hypersonic drones is pushing the limits of what's possible. New materials with enhanced thermal resistance, improved ablation characteristics, and better structural properties are constantly being developed and tested. Advancements in computational modelling and simulation techniques allow for more accurate predictions of the thermal environment and more efficient designs. Improved ground-based testing capabilities enable rigorous evaluation of the performance of materials and components under extreme conditions. Finally, ongoing flight testing provides invaluable data for validating these advancements and improving the technology even further. The continuous interplay between materials science, thermal management, and advanced computational techniques is crucial to unlocking the full potential of hypersonic drone technology. This field holds significant promise for both military and civilian applications, potentially revolutionizing high-speed transportation, aerospace exploration, and numerous other fields.

Maintaining stability and precision at these speeds requires sophisticated navigation and control systems capable of operating under intense aerodynamic forces and potentially harsh environmental conditions. The challenges are multifaceted and demand a robust integration of multiple technologies.

One of the primary challenges is the drastically reduced reaction time available for corrections. At hypersonic speeds, even minor deviations from the intended trajectory can quickly escalate into significant errors. This necessitates extremely fast and precise actuators and control algorithms capable of responding nearly instantaneously to sensor feedback. Traditional control systems, optimized for slower speeds, are simply inadequate. The dynamics of hypersonic flight are far more complex, involving highly nonlinear aerodynamic effects and significant coupling between different degrees of freedom. Therefore, sophisticated control algorithms, often relying on advanced mathematical models and real-time computations, are essential.

Control algorithms used in hypersonic drone navigation often employ advanced techniques like model predictive control (MPC) and nonlinear control methods. MPC is a particularly useful technique because it predicts the future behaviour of the system based on a dynamic model and then optimizes the control inputs to achieve the desired trajectory while respecting constraints on actuator limits and system dynamics. This allows for proactive control adjustments, anticipating potential deviations before they become significant problems. Nonlinear control methods are necessary because the aerodynamic forces acting on the drone at hypersonic speeds are highly nonlinear functions of its attitude, velocity, and altitude. These techniques incorporate advanced mathematical tools to effectively manage this complexity.

Another key element of hypersonic drone navigation is the reliance on highly accurate and robust sensor systems. Traditional GPS systems, while reliable at lower speeds, become less effective at hypersonic speeds due to the increased susceptibility to atmospheric interference and the possibility of jamming. Therefore, alternative or complementary navigation techniques are crucial. Inertial Navigation Systems (INS), utilizing accelerometers and gyroscopes to measure the drone's acceleration and angular rates, play a vital role. INS provides continuous position and attitude information, but they are prone to drift over time. This drift must be corrected using other sensor data.

This is where other sensor technologies such as star trackers, and advanced radar altimeters, become important. Star trackers can provide highly accurate celestial navigation information, independent of ground-based systems and relatively immune to atmospheric effects. These systems utilize sophisticated image processing algorithms to identify and track stars, providing precise heading and attitude references. Radar altimeters offer another layer of independent altitude measurement, vital for maintaining safe flight profiles. The fusion of data from multiple sensor systems – INS, GPS (when available), star trackers, radar altimeters, and potentially others – is essential for robust and reliable navigation.

This sensor data fusion is achieved using advanced estimation algorithms such as Kalman filtering or its variants. These algorithms combine data from various sensors, accounting for their inherent uncertainties and biases, to generate a more accurate and reliable estimate of the drone's state (position, velocity, attitude, etc.). The sophisticated algorithms also take into account the inherent uncertainties and noise present in the sensor data, resulting in a more reliable overall navigation solution. The development of these algorithms is an ongoing process, constantly refined to account for the complexities of hypersonic flight and the ever-improving accuracy and availability of sensor technologies.

Furthermore, the control systems must be designed to handle potential sensor failures. A robust system must be able to gracefully degrade its performance even if one or more sensors malfunction. This requires redundant sensors and intelligent fault-tolerant control algorithms capable of adapting to sensor failures in real-time. These algorithms continuously monitor the health of the sensor systems and employ appropriate fall-back strategies in case of failure. For example, if the GPS signal is lost, the system might rely more heavily on INS and star tracker data, or adjust the flight plan to compensate.

The computational demands of these advanced control algorithms are substantial. Hypersonic drones require high-performance onboard computers capable of processing vast amounts of sensor data and executing complex control calculations in real-time. The design of these systems requires careful consideration of processing power, memory capacity, and power consumption. The computing architecture must be resilient to the extreme thermal and vibration stresses of hypersonic flight, requiring specialized hardware components. Furthermore, efficient software algorithms are crucial to optimize performance and minimize computational latency. This necessitates the use of specialized programming languages and parallelization techniques to maximize computational efficiency.

Another significant challenge is the need for high-bandwidth communication links, particularly for drones operating beyond line-of-sight. Data from remote sensors and control commands from ground stations must be transmitted and received reliably at high data rates. This requires advanced communication technologies such as high-frequency data links and potentially satellite-based communication systems. These systems must be designed to be robust against interference and jamming, ensuring reliable communication in challenging environments. The high-speed data transfer is essential for both the real-time control of the drone and for transmitting telemetry data back to the ground station, enabling monitoring and analysis of flight parameters.

The extreme aerodynamic forces and thermal stresses associated with hypersonic flight impose significant demands on the drone's actuators. These actuators, responsible for controlling the flight surfaces and thrust vectoring, must be capable of operating reliably under these extreme conditions. This necessitates the development of advanced actuators with high strength-to-weight ratios, exceptional thermal resistance, and fast response times. The design of these actuators often involves the use of advanced materials and specialized manufacturing techniques. Furthermore, the control algorithms must be designed to accommodate the limitations and characteristics of these actuators, preventing damage and ensuring precise control.

The development of navigation and control systems for hypersonic drones is an area of intense ongoing research and development. The challenges are significant, but the potential benefits are equally compelling. The advancements in this field will pave the way for a new generation of high-speed, long-range drones with capabilities far exceeding those currently available. The continuous innovation in sensor technology, control algorithms, and onboard computing power is decisive for unlocking the full potential of hypersonic drone technology. This is not only essential for military applications, but it also has important implications for civilian applications, such as high-speed transportation and atmospheric research. The interplay between these advancements will undoubtedly reshape the future landscape of high-speed flight.

Building upon the challenges of navigation and control at hypersonic speeds, the critical aspect of survivability necessitates the integration of advanced stealth technologies and evasive manoeuvres. The extreme speed of these drones, while offering a significant advantage in terms of rapid deployment and strike capabilities, also presents a unique vulnerability: detection and engagement by enemy air defence systems. Therefore, minimizing the probability of detection and maximizing the chances of evasion become paramount considerations in the design and operation of hypersonic drones.

One of the primary methods for enhancing survivability is the reduction of the drone's radar cross-section (RCS). RCS is a measure of the strength of the radar signal reflected by an object. A smaller RCS makes the drone harder to detect by radar systems. This is achieved through a combination of design features and material science advancements.

Shape optimization plays a fundamental role in RCS reduction. Hypersonic drones are often designed with features like sharp edges, angled surfaces, and RAM (Radar Absorbent Material) coatings. These design choices disrupt the reflection of radar waves, scattering them in multiple directions and reducing the intensity of the reflected signal detected by the radar. The use of curved surfaces, particularly at the leading edges, can further enhance stealth capabilities by reducing specular reflections — direct reflections of radar waves back towards the source.

The integration of RAM is another vital aspect of stealth technology. RAM materials are specifically designed to absorb radar energy, thus minimizing the reflected signal. Different types of RAM exist, each with its own characteristics and applications. Some RAM materials are designed to absorb a broad range of radar frequencies, while others are optimized for specific frequency bands. The choice of RAM material is often determined by the specific threat environment and the desired level of stealth. Furthermore, the application of RAM is not trivial and can be constrained by factors such as the high temperatures associated with hypersonic flight. This necessitates the development of high-temperature RAM materials capable of withstanding the extreme thermal stresses of hypersonic flight without compromising their radar-absorbing properties.

Beyond material science, the careful selection of surface coatings is another critical factor in RCS reduction. These coatings can minimize reflections by either absorbing radar waves or by scattering them in less detectable directions. This is achieved through the controlled manipulation of the surface impedance, affecting the reflection and absorption characteristics of the surface with respect to electromagnetic waves. In some cases, specially designed coatings act as wave cancellers, interfering with the reflected signals to diminish the overall RCS.

Beyond passive stealth measures, active countermeasures are also being investigated. These could involve deploying counter-radar chaff or electronic jamming systems to interfere with enemy radar systems, making it difficult to acquire and track the hypersonic drone. Chaff dispensers release metallic strips or fibbers which reflect radar signals, creating a false target cloud and masking the actual drone. Electronic jamming actively disrupts the radar signals, creating noise and interfering with the target acquisition process. These systems require careful design and integration to ensure they do not compromise the drone's other critical systems and to function effectively in high-speed and high-g environments. Moreover, their use often involves trade-offs between effectiveness and detectability.

Evasive manoeuvres are another important element of survivability. At hypersonic speeds, even small deviations from the planned trajectory can significantly alter the engagement geometry, making it more challenging for enemy systems to track and intercept. These manoeuvres require sophisticated control algorithms capable of generating and executing rapid trajectory adjustments based on sensor information and threat assessments. Precise flight control is paramount to ensure manoeuvrability without causing structural damage due to the intense aerodynamic forces.

The use of AI plays a crucial role in optimizing these evasive manoeuvres. Autonomous systems can process sensor data in real-time and react far more quickly than human operators could. AI algorithms can analyze radar and other sensor data, identifying potential threats and calculating optimal trajectories to evade detection and engagement. Machine learning techniques are also being applied to improve the effectiveness of evasive manoeuvres over time. By analyzing past encounters with air defence systems, AI algorithms can identify patterns and improve their ability to predict and counteract enemy actions.

Advanced sensor fusion is also vital to the successful implementation of evasive manoeuvres. Hypersonic drones often rely on a suite of sensors to monitor their surroundings and detect potential threats. This data is combined using sophisticated algorithms to create a comprehensive picture of the threat environment. The algorithms then determine the optimal evasive manoeuvres, balancing the need to avoid detection with the need to maintain trajectory and mission objectives. Incorporating data from various sensors, such as infrared cameras, passive and active radar systems, and other electronic warfare (EW) sensors is vital in generating a comprehensive and robust situational awareness picture. This enables the autonomous system to swiftly adapt to changing environments and optimize the avoidance strategies in real-time.

The development of hypersonic drones with enhanced stealth capabilities and sophisticated evasive manoeuvres is an area of intense research and development. The challenges are significant but the potential rewards are equally substantial. By combining advanced materials, innovative design features, and sophisticated AI algorithms, engineers are working towards creating drones that are significantly more difficult to detect and intercept, enabling them to operate effectively in contested airspace. The continuous integration of advancements in materials science, control systems, and AI will undoubtedly lead to even more effective stealth and evasion technologies in the future. Moreover, the development and refinement of these technologies will also impact civilian applications, such as high-speed transportation and satellite deployment, which necessitate the ability to manoeuvre safely through adverse conditions. The balance between achieving maximum speed and survivability remains an area of ongoing engineering optimization, aiming for fast, effective, and difficult-to-detect systems.

The development of hypersonic drones presents a paradigm shift in military capabilities, offering unprecedented speed and potential for bypassing established air defence systems. Their ability to rapidly deploy and strike targets with precision introduces a new dimension to military strategy and tactics. The implications extend far beyond simply increasing the speed of existing drone technology; they represent a fundamental alteration in the balance of power and the nature of conflict.

One of the most significant military applications of hypersonic drones is their potential for conducting rapid precision strikes against high-value targets. Traditional air-launched cruise missiles, while possessing long range, are comparatively slow, giving enemy defences ample time to react. Hypersonic drones, however, can significantly reduce this reaction time. Their immense speed allows them to penetrate enemy airspace quickly, deliver their payload, and potentially evade interception before defensive systems can be effectively engaged. This speed advantage is particularly crucial in scenarios where time-sensitive targets, such as mobile missile launchers or command and control centres, are involved. The reduced exposure time minimizes the risk of the drone being detected and destroyed, enhancing mission success rates.

Furthermore, the speed of hypersonic drones contributes to their effectiveness against geographically dispersed targets. In scenarios requiring rapid response to unfolding events across vast distances, the speed advantage of these drones becomes a game-changer. They could potentially be deployed from various locations, such as ships, submarines, or land-based mobile launchers, providing a flexible and responsive strike capability. This capability is highly valuable for addressing time-critical threats, rapidly responding to contingencies, and potentially disrupting enemy operations before they can fully materialize. The swiftness of deployment and engagement offered by hypersonic drones makes them an effective deterrent, potentially dissuading adversaries from escalating conflicts.

Another key advantage lies in their potential to bypass existing air defence systems. Traditional air defence systems are designed to engage threats at subsonic and supersonic speeds. The extreme speed of hypersonic drones can overwhelm these systems, making interception extremely difficult. The combination of high speed and advanced manoeuvrability makes these drones exceptionally challenging to track and engage, even with the most sophisticated radar and missile systems. The manoeuvring capabilities, often coupled with advanced countermeasures, enhance survivability and increase the chances of successful mission completion. The current generation of air defence systems may be rendered largely ineffective against a concerted attack utilizing hypersonic drones.

The strategic implications of hypersonic drone technology are profound. The ability to deliver precision strikes rapidly and with a high degree of survivability can fundamentally alter the calculus of military engagements. States possessing such capabilities gain a significant advantage in asymmetric warfare scenarios, where opponents may possess superior numbers but lack the ability to rapidly counter hypersonic attacks. The rapid deployment capability also affects the strategic decision-making processes, enabling faster and more decisive responses to developing situations. This might lead to a shift in the nature of deterrence, with nations relying more on the threat of rapid and decisive strikes rather than maintaining large standing armies or relying on long-term deployments.

Potential mission profiles for hypersonic drones are diverse and encompass a wide range of military applications. They could be employed for pre-emptive strikes against enemy infrastructure, targeting critical assets such as command centres, missile silos, or airfields. Their ability to quickly neutralize these assets significantly reduces an enemy's ability to respond to subsequent attacks. Hypersonic drones could also play a pivotal role in counter-terrorism operations, targeting high-value individuals or terrorist cells in remote or heavily defended locations. Their speed and precision make them ideal for minimizing civilian casualties while maximizing operational effectiveness. Beyond conventional military applications, these drones could be instrumental in special operations, delivering critical supplies or providing close air support to ground forces operating in hostile environments.

However, the development and deployment of hypersonic drones also raise significant challenges. The extreme speeds involved place immense stress on the drone's structural integrity and require the use of advanced materials and manufacturing techniques. The high temperatures generated during hypersonic flight pose a significant challenge to the design and operation of these systems. Furthermore, integrating advanced sensors, navigation systems, and sophisticated AI algorithms within a hypersonic platform requires significant technological advancements. The cost of developing and deploying such advanced systems is also a considerable factor, potentially limiting their widespread adoption, at least in the near term.

The development of effective countermeasures is another crucial area. While hypersonic drones present significant challenges to current air defence systems, ongoing research and development efforts are focused on developing new technologies to counter them. This involves improving existing radar systems to better detect and track hypersonic targets, developing new types of missiles capable of intercepting them, and exploring advanced electronic warfare techniques to disrupt their operations. The arms race triggered by the development of hypersonic drones is likely to continue, with both offensive and defensive technologies constantly evolving.

The ethical and legal implications of hypersonic drones also demand careful consideration. Their speed and precision raise concerns about the potential for accidental or unintended casualties. The use of autonomous systems in lethal applications raises complex questions about accountability and the potential for unintended escalation. International cooperation and the development of appropriate legal frameworks are crucial to ensure the responsible development and deployment of this technology. The international community needs to establish norms and guidelines to prevent the proliferation of this technology and to minimize the risks of unintended consequences.

The Potential of Drone Technology

The integration of artificial intelligence (AI) and machine learning (ML) is poised to revolutionize drone technology, pushing the boundaries of what's possible in both military and civilian applications. The advancements being made are not simply incremental improvements; they represent a fundamental shift in the capabilities and operational potential of unmanned aerial vehicles (UAVs). We are moving beyond pre-programmed flight paths and basic obstacle avoidance towards a future where drones exhibit a level of autonomy and adaptability previously confined to the realm of science fiction.

One of the most significant areas of advancement lies in enhanced autonomy. Current drones, even the most sophisticated ones, often require significant human intervention, particularly in complex or unpredictable environments. Future AI-powered drones will possess far greater decision-making capabilities, enabling them to navigate challenging terrain, avoid obstacles autonomously, and adapt to changing conditions in real-time without constant human oversight. This enhanced autonomy will be achieved through advancements in several key areas.

Firstly, improvements in computer vision and image recognition will allow drones to "see" and interpret their surroundings with far greater accuracy and detail than is currently possible. This involves utilizing advanced deep learning algorithms trained on massive datasets of imagery, enabling them to identify objects, assess distances, and understand the context of their environment with a speed and accuracy surpassing human capabilities. This will greatly improve situational awareness, allowing drones to make informed decisions regarding navigation, target acquisition, and threat avoidance. For instance, a drone could autonomously navigate a dense urban environment, identify and avoid obstacles like power lines, buildings, and pedestrians, all without human input. This capability is critical for both civilian and military applications, improving efficiency and reducing the risk of accidents or mission failure.

Secondly, advancements in reinforcement learning will enable drones to learn from their experiences and refine their decision-making processes over time. Reinforcement learning algorithms allow the drone to learn optimal behaviours through trial and error, adapting its strategies based on feedback from the environment. This is particularly important for tasks requiring adaptability and improvisation, such as search and rescue operations in disaster zones or complex military engagements. Imagine a drone deployed in a wildfire, learning to navigate through shifting smoke, avoid falling trees, and efficiently locate and rescue trapped individuals. The ability to learn and adapt in such dynamic environments would drastically improve the effectiveness of such missions.

Moreover, improved path planning algorithms, driven by AI, will lead to more efficient and effective drone operations. Currently, drone flight paths are often pre-programmed, limiting their ability to respond to unforeseen circumstances. Future AI-powered path planning algorithms will allow drones to dynamically adjust their routes based on real-time information, optimizing for speed, fuel efficiency, and obstacle avoidance. This is particularly important for long-range missions, where even small improvements in efficiency can significantly extend operational range and endurance. For military applications, this means drones could autonomously adjust their flight paths to avoid enemy defenses or take advantage of unexpected opportunities.

The application of AI and ML extends beyond autonomous navigation. It plays a critical role in enhancing target recognition and engagement capabilities. Drones equipped with advanced

AI-powered image recognition systems can identify and classify targets with remarkable speed and accuracy. This can improve the precision of strikes, minimizing collateral damage and reducing the risk of civilian casualties. Furthermore, AI algorithms can be used to analyze vast amounts of sensor data in real-time, identifying patterns and anomalies that might otherwise go unnoticed by human operators. This enhanced situational awareness can provide critical intelligence for both military commanders and civilian emergency responders.

The development of swarm technology, enabled by AI, represents another significant leap forward. Swarms of drones coordinated and controlled by sophisticated AI algorithms, can collectively accomplish tasks that would be impossible for a single drone to handle. These swarms can collaboratively map vast areas, conduct coordinated searches, or even launch coordinated attacks against multiple targets. The decentralized nature of swarm intelligence makes them highly resilient to loss of individual drones and capable of adapting to changing circumstances. The implications for military operations are significant, offering the potential for overwhelming firepower and adaptable strategies that can overwhelm conventional defences. Moreover, swarm technology is not limited to military applications; it has vast potential in civilian sectors, such as environmental monitoring, infrastructure inspection, and search and rescue operations.

However, the advancements in AI and ML for drone technology are not without challenges. One major concern is the ethical implications of increasingly autonomous weapon systems. As drones become capable of independent decision-making regarding target selection and engagement, the question of accountability becomes critical. The potential for unintended consequences, accidental killings, and the erosion of human control over lethal force requires careful consideration. Establishing clear guidelines and ethical frameworks is crucial to ensure responsible development and deployment of these technologies.

Another challenge is the vulnerability of AI systems to cyber attacks. If an adversary is able to compromise the AI controlling a drone, it could be repurposed for malicious purposes, potentially with devastating consequences. Therefore, robust cyber security measures are crucial to protect these systems from hacking and manipulation. This includes developing secure communication protocols, employing robust encryption techniques, and regularly updating software to patch vulnerabilities. Moreover, research into AI-resistant countermeasures is necessary to protect against attacks that try to mislead or disable AI-driven systems.

Furthermore, the development and deployment of AI-powered drones require significant computational resources and sophisticated software engineering. Training advanced AI algorithms requires vast amounts of data and powerful computing infrastructure. This presents a significant technological challenge, particularly for smaller nations or organizations with limited resources. The disparity in technological capabilities could further exacerbate existing inequalities in global power dynamics.

The effectiveness of any autonomous system hinges critically on its ability to accurately perceive and interpret its environment. This necessitates a leap forward in sensor technology and the sophisticated processing of the data these sensors collect. The next generation of drones will rely on a diverse array of sensors working in concert, fused together through advanced algorithms to provide an unparalleled level of situational awareness.

One of the most significant advancements is the miniaturization and increased sensitivity of existing sensor technologies. High-resolution cameras, already common on many drones, are undergoing rapid improvement, offering higher resolutions, wider fields of view, and improved low-light performance. This enhanced visual acuity allows for more accurate object identification and tracking, crucial for both civilian and military applications. For instance, improved cameras can enable more precise agricultural monitoring by identifying individual plants exhibiting stress or disease. In a military context, enhanced resolution can differentiate between combatants and civilians, minimizing collateral damage.

Beyond visual sensors, infrared (IR) cameras are becoming increasingly sophisticated. These sensors detect heat signatures, providing valuable information even in low-light or obscured conditions. This is particularly relevant for search and rescue operations, where locating individuals in smoke-filled environments or at night is paramount. Similarly, IR cameras are valuable in military applications for detecting camouflaged targets or tracking enemy movement. Advancements in IR technology are leading to smaller, lighter, and more energy-efficient sensors, making them more suitable for integration onto smaller and more agile drones.

LiDAR (Light Detection and Ranging) technology is another crucial area of advancement. LiDAR sensors use lasers to create detailed 3D maps of the surrounding environment, providing highly accurate distance measurements and object identification. This technology is proving particularly useful in autonomous navigation, enabling drones to navigate complex terrains, avoid obstacles effectively, and create accurate representations of their surroundings for mapping or inspection purposes. LiDAR is already finding applications in construction, surveying, and precision agriculture, and its integration into smaller and more affordable drones will further expand its applicability.

Radar systems, often used in larger UAVs, are also experiencing significant advancements. Smaller, more energy-efficient radar systems are emerging, capable of detecting objects at greater distances and with higher accuracy. This is particularly important for applications requiring long-range surveillance or target tracking in challenging environments, such as fog or rain. The integration of radar with other sensors, like cameras and LiDAR, can provide a more complete and robust understanding of the drone's environment, even in conditions where visibility is severely limited.

Beyond these established sensor technologies, newer and more experimental sensors are emerging. Hyper spectral imaging, for example, captures images across a much broader range of wavelengths than conventional cameras, allowing for the identification of materials based on their unique spectral signatures. This could revolutionize agricultural monitoring, allowing for the precise detection of nutrient deficiencies or disease in plants. In a military context, hyper spectral imaging could identify hidden explosives or camouflage materials.

The development of multispectral cameras, capable of capturing images in different parts of the electromagnetic spectrum, simultaneously, is another promising area. These cameras can fuse data from various spectral bands to provide enhanced image quality and information. For example, they can combine visible light with near-infrared to improve vegetation analysis and crop monitoring. In the military sphere, multispectral cameras could enhance target identification and recognition, particularly in challenging conditions.

However, having access to a vast amount of sensor data is only half the battle. Effectively processing and integrating this information to create a comprehensive situational awareness picture requires advanced data fusion techniques. These techniques combine information from multiple sensors to create a more complete and accurate representation of the environment than any single sensor could provide. Advanced algorithms, often based on artificial intelligence and machine learning, are used to correlate and interpret the data, removing redundancies and highlighting inconsistencies.

Data fusion algorithms are becoming increasingly sophisticated, capable of handling large volumes of data from diverse sensor sources in real-time. These algorithms employ techniques like Kalman filtering and Bayesian inference to estimate the most likely state of the environment, taking into account the uncertainty associated with each sensor reading. The output of these algorithms is a highly accurate and reliable representation of the drone's surroundings, including the location, orientation, and characteristics of objects within its field of view.

The development of robust data fusion techniques is crucial for enabling greater autonomy in drones. By integrating information from multiple sensors, drones can develop a more complete understanding of their environment, making more informed decisions regarding navigation, obstacle avoidance, and target acquisition. For instance, a drone might use LiDAR data to create a 3D map of its surroundings, then combine that information with camera data to identify and classify objects within that map, thereby improving navigation and target recognition. This integrated approach is critical for enhancing the reliability and safety of autonomous drone operations.

Furthermore, the advancements in data fusion are driving the development of more robust and resilient drone systems. By integrating information from multiple sensors, drones can compensate for sensor failures or environmental limitations. If one sensor fails, the data fusion algorithm can still provide a reliable situational awareness picture by relying on the information from other sensors. This redundancy is crucial for maintaining the operation of critical systems even in the face of unexpected events or adversarial attacks.

The improved sensor technology and data fusion techniques are not simply enhancing drone capabilities incrementally; they are fundamentally changing how drones operate and interact with their environments. This represents a paradigm shift in autonomous systems, paving the way for more reliable, adaptable, and efficient drones across diverse applications. Future drones will be far more than simple flying robots; they will be sophisticated intelligent agents capable of perceiving, understanding, and interacting with the world around them in unprecedented ways.

The development and deployment of these advanced sensor and data fusion systems will require significant advancements in both hardware and software. This includes the development of more powerful and energy-efficient processors capable of handling the immense computational demands of real-time data processing. Moreover, the development of advanced algorithms capable of efficiently fusing data from diverse sensor sources is essential. The progress in these areas is crucial for realizing the full potential of next-generation drone technology. The future of drone technology lies not just in the individual capabilities of the sensors themselves but in the sophisticated integration and intelligent processing of the vast quantities of information they provide. This synergy between sensor advancements and data fusion techniques represents the key to unlocking truly autonomous and intelligent unmanned aerial vehicles.

Integration Technologies – Air – Land and Sea

The true potential of unmanned aerial vehicles (UAVs) may lie not solely in their individual capabilities, but in their seamless integration with other autonomous systems. This integration promises to create a synergistic effect, generating capabilities far exceeding the sum of their individual parts. The future battlefield, and indeed many civilian applications, will likely be characterized by a coordinated network of autonomous systems working in concert, with drones playing a crucial, yet interconnected, role.

One compelling example is the integration of drones with ground robots. Imagine a scenario involving search and rescue operations in a disaster zone. A fleet of small, agile drones could be deployed to rapidly survey the affected area, providing real-time aerial imagers and identifying potential survivors or obstacles. This aerial reconnaissance data could then be relayed to a network of ground robots equipped with sensors and manipulation capabilities. The robots, guided by the drone's information, could navigate the debris field, access hard-to-reach areas, and deliver essential supplies or provide immediate medical assistance. This combined aerial and ground approach could significantly enhance the speed and efficiency of rescue operations, potentially saving lives and minimizing further damage.

The collaboration between drones and ground robots extends beyond disaster relief. In military contexts, this integration can drastically improve situational awareness and operational effectiveness. Drones can provide long-range surveillance, detecting enemy positions and movements, transmitting this crucial information to ground robots positioned closer to the target. The ground robots, in turn, might deploy sensors for more detailed reconnaissance or even engage in direct action, guided by the real-time intelligence provided by the drones. This coordinated effort reduces human risk while significantly enhancing the speed and accuracy of military operations. For example, a drone could identify a concealed enemy position, while a ground robot equipped with a small weapon system could then engage the target, minimizing friendly casualties and maximizing operational success.

The integration of drones with autonomous vehicles, such as self-driving cars and trucks, opens up yet another exciting area of collaborative autonomy. Consider the use of autonomous delivery systems. Drones could serve as the "last mile" solution, delivering packages from a central distribution point to individual addresses. Autonomous vehicles could transport the packages from larger warehouses to designated drop-off points, where drones could then take over, completing the final leg of the delivery journey. This combined system could significantly enhance the speed and efficiency of delivery services, reducing costs and expanding the reach of e-commerce to even the most remote areas. Furthermore, this synergistic approach could also improve safety, as the autonomous vehicle and drone system would operate under controlled protocols, reducing the risks associated with human-driven transportation.

However, the integration of drones with other autonomous systems presents significant technical challenges. One of the most critical challenges is ensuring seamless communication and data sharing between these different systems. This requires the development of robust communication protocols and standardized data formats to guarantee efficient and reliable information exchange. Different autonomous systems often operate using various communication protocols and data structures, leading to interoperability problems. Addressing this through standardized protocols and data models is essential for successful integration.

Another critical challenge is the development of robust algorithms for coordinated control. These algorithms must seamlessly manage the interactions between different systems, ensuring that they work together harmoniously to achieve a common objective. This task necessitates advanced AI algorithms that can handle complex decision-making processes and adapt to dynamic operational environments. The algorithms must also account for potential failures or unexpected events in any of the participating systems, ensuring that the overall mission remains resilient and can adapt to changing conditions.

Furthermore, the issue of security is vital when integrating various autonomous systems. Cyber security threats become more complex as the number of interconnected systems increases, making the entire network vulnerable to potential hacks or malicious attacks. Developing robust cyber security measures to protect the integrity of the integrated system becomes absolutely critical, especially in military or sensitive civilian applications.

Despite these challenges, the potential benefits of integrating drones with other autonomous systems are immense. The combination of aerial and ground perspectives, coupled with the adaptability and responsiveness of AI-driven control systems, opens up a wide array of new possibilities across various sectors. From enhancing emergency response capabilities to revolutionizing transportation and logistics, the synergy between drones and other autonomous systems is transforming the landscape of multiple industries. Further research and development in areas such as communication protocols, coordinated control algorithms, and cyber security measures are essential to unlocking the full potential of this transformative technology, paving the way for a future where autonomous systems work collaboratively to address some of the most challenging problems facing humanity. The next frontier of drone technology is not just in improving the capabilities of individual drones, but in harnessing their collective potential through seamless integration with a broader ecosystem of autonomous systems. The challenges are significant, but the rewards are potentially transformative.

The future of drone technology is inextricably linked to the development and refinement of human-machine teaming (HMT) and collaborative control strategies. These strategies are not about replacing human operators but about optimizing their interaction with increasingly sophisticated drones for enhanced performance, safety, and ethical considerations.

The traditional model of drone operation, where a human pilot directly controls every aspect of the drone's flight and mission, is rapidly becoming obsolete. As drones gain greater autonomy, the role of the human operator shifts from direct control to oversight and strategic decision-making. This necessitates a fundamental change in the design of human-machine interfaces (HMIs), moving away from complex manual control systems towards intuitive and user-friendly interfaces that facilitate seamless collaboration.

Effective HMI requires a clear understanding of the strengths and limitations of both humans and machines. Humans excel at complex reasoning, adaptability, and ethical judgment, while machines are superior in processing vast amounts of data quickly and accurately, executing repetitive tasks reliably, and operating in hazardous environments. Optimal HMI leverages these complementary capabilities, allowing human operators to focus on higher-level tasks that require judgment and strategic thinking, while the drones handle the more mundane, data-intensive aspects of the mission.

Modern HMIs are moving beyond traditional joystick-based control systems and are incorporating advanced technologies such as augmented reality (AR) overlays, virtual reality (VR) simulations, and natural language processing (NLP). AR overlays can project crucial flight data directly onto the pilot's view, providing real-time situational awareness without diverting attention from the visual environment. VR simulations allow pilots to train in realistic virtual environments, preparing them for various operational scenarios and improving their decision-making skills in complex situations. NLP interfaces enable more natural interaction between pilots and drones, allowing for voice commands and conversational interactions, reducing the cognitive load on the operator and improving overall efficiency.

The design of HMIs also needs to consider the context of operation. The interface for a civilian drone used for aerial photography will differ significantly from the HMI for a military drone involved in complex combat scenarios. For example, a civilian drone HMI might prioritize ease of use and intuitive controls, while a military HMI might emphasize information density, real-time threat assessment, and seamless integration with other battlefield systems.

Collaborative control strategies go beyond simply improving the HMI. They involve the development of algorithms that enable seamless sharing of control authority between human operators and autonomous systems. This shared control approach allows for a dynamic balance between human intervention and autonomous operation, adapting to changing circumstances and optimizing performance based on the specific situation.

The drones, equipped with advanced AI algorithms for obstacle avoidance and target recognition, would then autonomously navigate the search areas, collecting data and relaying it back to the operator. The operator could then intervene if necessary, directing the drones to specific locations or adjusting search parameters based on the information received. This shared control approach enables efficient and effective search operations, optimizing the use of human expertise and machine capabilities.

Similarly, in military applications, collaborative control can be vital for complex missions requiring real-time decision-making. A human operator might oversee a swarm of drones conducting a reconnaissance mission, providing overall strategic guidance while allowing the drones to autonomously navigate, avoid obstacles, and detect potential threats. The operator could intervene if a situation requires human judgment, overriding autonomous decisions or redirecting the swarm's activity based on evolving circumstances. This collaborative approach improves operational effectiveness, reduces human risk, and allows for rapid adaptation to dynamic battlefield conditions.

The transition from air-based drone warfare to the maritime domain introduces a new set of challenges and opportunities. While aerial drones dominate headlines, the silent, persistent operation of unmanned surface vessels (USVs) is quietly revolutionizing naval capabilities, particularly in surveillance and mine countermeasures (MCM). USVs offer a compelling blend of affordability, endurance, and adaptability, making them a crucial element of modern naval strategy. Unlike their airborne counterparts, USVs operate within the complex three-dimensional environment of the ocean, necessitating robust design features and sophisticated control systems to navigate currents, waves, and unpredictable weather conditions.

One of the primary applications of USVs is maritime surveillance. Their ability to patrol vast stretches of ocean for extended periods, often without the need for a human crew onboard, provides a cost-effective solution for monitoring maritime borders, protecting critical infrastructure, and detecting illegal activities such as smuggling, piracy, and illegal fishing. Equipped with a range of sensors, including radar, electro-optical/infrared (EO/IR) cameras, and acoustic sensors, USVs can provide persistent surveillance, collecting real-time intelligence on maritime traffic, weather patterns, and potential threats. The data collected by these sensors can be transmitted back to a central command centre, allowing for real-time monitoring and analysis. This persistent surveillance capability frees up larger, more expensive naval assets for other missions, improving overall operational efficiency.

The endurance of USVs is a significant advantage in maritime surveillance. Unlike manned vessels, which require regular crew changes and resupply, USVs can operate autonomously for weeks or even months, depending on their design and size. This extended endurance enables them to maintain a continuous presence in remote or challenging areas, providing crucial intelligence gathering capabilities that would be impractical or impossible to achieve with manned vessels. For example, a USV could be deployed to monitor a strategically important shipping lane for extended periods, detecting and tracking any suspicious activity. The data collected could then be used to inform decision-making by naval commanders, allowing for a more proactive and effective response to potential threats.

The relatively low cost of USVs compared to manned vessels makes them an attractive option for navies with limited budgets. This affordability allows for the deployment of larger numbers of USVs, creating a network of surveillance platforms that can cover a much wider area. Furthermore, the modular design of many USVs allows for customization based on specific mission requirements. For example, a USV designed for maritime surveillance could be easily reconfigured to carry different payloads, such as specialized sensors or communication equipment, adapting to evolving operational needs. This flexibility significantly enhances their versatility and operational value.

Beyond surveillance, USVs are proving increasingly valuable in mine countermeasures (MCM). Mines remain a significant threat to naval operations, and traditional MCM methods are often time-consuming, labour-intensive, and dangerous. USVs offer a safer and more efficient alternative. Equipped with advanced sonar systems and robotic manipulators, USVs can be deployed to detect, classify, and even neutralize mines without putting human divers or specialized MCM vessels at risk. The autonomous operation of USVs allows for the rapid clearance of minefields, reducing the time required for naval forces to transit through potentially hazardous areas. This enhances operational readiness and reduces the risk to personnel and valuable assets.

The use of autonomous underwater vehicles (AUVs) in conjunction with USVs significantly enhances MCM capabilities. AUVs can be deployed from USVs to conduct detailed underwater surveys, providing high-resolution imagery and data about the seafloor, helping to pinpoint the location of mines. The AUVs can then relay this information back to the USV, allowing for a more precise and effective mine neutralization strategy. This collaborative approach combines the endurance and persistent presence of USVs with the detailed underwater survey capabilities of AUVs, creating a highly effective MCM system. This synergistic relationship showcases the evolving capabilities of unmanned systems working in concert.

Different USV platforms cater to specific operational needs. Smaller, more agile USVs might be ideal for coastal surveillance and shallow-water operations, while larger, more robust platforms are better suited for open-ocean missions. The payload capacity of a USV is another crucial factor. Larger USVs can carry more sensors, communication equipment, and potentially weapons systems, enhancing their operational capabilities. However, this increased payload capacity often comes at the expense of speed and manoeuvrability. The design and operational parameters of a USV must be carefully considered based on its intended mission profile and the operational environment.

Environmental factors also play a significant role in the effectiveness of USVs. Harsh weather conditions, such as high waves and strong currents, can limit the operational capabilities of smaller USVs. Larger, more robust platforms are generally more resilient to these challenges, but even these can face limitations in extreme weather. The design of the USV hull, its propulsion system, and its onboard control systems must be robust enough to withstand the rigors of the marine environment. The development of advanced autonomous navigation systems that can compensate for challenging environmental conditions is crucial to improve the reliability and operational effectiveness of USVs.

The integration of artificial intelligence (AI) is further enhancing the capabilities of USVs. AI-powered systems can improve the efficiency of sensor data analysis, enhancing the detection of threats and reducing the workload on human operators.

AI can also improve autonomous navigation, allowing USVs to navigate complex environments more effectively. The development of more sophisticated AI algorithms will undoubtedly lead to increased autonomy and improved operational capabilities for USVs in the future. The ability of AI to learn and adapt from past experiences can improve the decision-making processes of USVs, making them more responsive to changing conditions.

Furthermore, the potential for swarming tactics with USVs is an area of active research and development. A coordinated group of USVs could overwhelm an opponent's defenses or conduct large-scale surveillance operations more effectively than a single platform. This coordinated action could enable more complex missions and a wider coverage area. However, managing and controlling a swarm of USVs requires sophisticated communication and control systems. The integration of secure communication protocols and robust command and control architectures is crucial for effective swarming operations. This requires significant advancements in both hardware and software to ensure reliable and coordinated actions.

The development and deployment of USVs are not without their challenges. The technological complexity of these systems requires significant investment in research and development. Ensuring the cyber security of USVs is also crucial, as these platforms could be vulnerable to cyber attacks. International regulations and norms governing the operation of USVs in international waters also need to be developed to prevent unintended conflicts and ensure responsible use. Addressing these challenges is vital for the safe and effective integration of USVs into naval operations.

The transition from surface unmanned systems to the underwater realm introduces a new level of complexity. While USVs operate on the surface, subject to the vagaries of weather and waves, unmanned underwater vehicles (UUVs) face an even more challenging environment. The crushing pressure, limited visibility, and unpredictable currents of the deep ocean demand robusts engineering and sophisticated control systems. Yet, the potential rewards – enhanced surveillance, reconnaissance, and mine countermeasures – are immense, driving significant investment and innovation in this field.

UUVs, unlike their surface counterparts, operate in a domain largely obscured from direct human observation. This necessitates reliance on advanced sensor technologies to gather intelligence and navigate. Sonar, in its various forms, is the cornerstone of UUV sensing. Side-scan sonar provides high-resolution images of the seafloor, revealing potential hazards or points of interest. Multibeam sonar generates a three-dimensional map of the seabed, crucial for navigation and detailed seabed analysis. Synthetic aperture sonar (SAS) offers even higher resolution than traditional sonar, capable of detecting smaller objects and subtle variations in the seafloor. These sonar systems, however, are not without their limitations. The range and resolution of sonar are often impacted by water conditions – turbidity, salinity, and temperature all affect the quality of the data received. Sophisticated signal processing algorithms are employed to mitigate these effects, but interpreting sonar data still requires considerable expertise.

Beyond sonar, UUVs often incorporate other sensors, such as magnetometers to detect metallic objects, like mines or submerged wreckage. Optical cameras, while limited in range due to water turbidity, can provide invaluable high-resolution imagery of targets of interest in relatively clear waters. Advanced UUVs may incorporate a combination of sensors, leveraging their complementary capabilities for a more comprehensive understanding of the underwater environment. Endurance is a critical factor in UUV operations. The energy requirements for propulsion, sensor operation, and communication place significant constraints on the operational time. Early UUVs were limited to relatively short deployments, but technological advancements, particularly in battery technology and energy-efficient propulsion systems, have significantly increased their endurance. Some modern UUVs can operate for weeks or even months without requiring recovery. This extended endurance allows for prolonged surveillance missions, monitoring critical infrastructure, or tracking long-range targets. However, the energy capacity still limits mission profiles, necessitating careful planning and efficient energy management strategies.

Communication with UUVs presents unique challenges. Radio waves are significantly attenuated by water, making reliable communication difficult or impossible beyond relatively short ranges. Acoustic communication is commonly employed, using sound waves to transmit data between the UUV and a surface vessel or shore station. However, acoustic communication is slower and less bandwidth-efficient than radio, limiting the amount of data that can be transmitted in a given timeframe. This necessitates efficient data compression and prioritization techniques to ensure that critical information is relayed reliably. The development of more efficient acoustic modems and alternative communication technologies, such as optical communication, is an active area of research, striving to increase both range and data throughput.

The autonomy of UUVs is constantly evolving. Early UUVs often required pre-programmed routes and relatively little real-time decision-making. Advances in artificial intelligence (AI) and machine learning are leading to increased autonomy, allowing UUVs to adapt to changing conditions and make real-time decisions based on sensor data. For example, an AI-powered UUV could autonomously navigate around obstacles, re-route around unexpected currents, or even prioritize targets based on their perceived threat level. This increased autonomy enhances the operational flexibility and effectiveness of UUVs, reducing reliance on human intervention, which in deep ocean or hazardous environments is highly challenging and risky.

Mine countermeasures (MCM) are a significant application for UUVs. Mines are a persistent threat to naval operations, capable of disrupting shipping lanes and damaging or sinking vessels. Traditional MCM methods can be time-consuming, dangerous, and costly. UUVs, equipped with advanced sonar and manipulators, provide a safer and more efficient approach. They can autonomously survey large areas, identify potential mines, and even neutralize them using various methods, depending on the type of mine encountered. The ability of UUVs to operate in shallow and deep waters greatly enhances their value in a wide range of operational scenarios.

The design and capabilities of UUVs vary significantly depending on their intended mission. Smaller, more agile UUVs are suitable for shallow-water operations, coastal surveillance, or inspections of ship hulls. Larger, more robust UUVs are capable of extended deployments in deep ocean environments, conducting larger-scale surveys or carrying heavier payloads. The payload capacity of UUVs is a significant design consideration, affecting their range, endurance, and the types of sensors and equipment they can carry. For example, a UUV designed for long-range surveillance might prioritize energy efficiency and endurance over payload capacity, while a UUV designed for mine neutralization might require a robust manipulator arm and sophisticated sonar, potentially at the expense of range.

The development and deployment of UUVs face several challenges. The harsh underwater environment demands robust and reliable systems capable of withstanding high pressures and corrosive saltwater. The cost of developing and manufacturing sophisticated UUVs remains high, but the cost-effectiveness compared to manned underwater operations is significant in the long term. Ensuring the cybersecurity of UUVs is crucial, as they could become targets for cyberattacks, potentially compromising sensitive information or disrupting their operation. International regulations governing the use of UUVs in international waters need further refinement to prevent conflict and ensure responsible use.

Technological advancements in areas such as battery technology, AI, and acoustic communication are continually driving improvements in UUV capabilities. The integration of AI is revolutionizing UUV autonomy, allowing for more complex missions and greater operational efficiency. The development of more robust and energy-efficient propulsion systems is extending operational endurance, enabling longer deployments and broader surveillance capabilities. Miniaturization of sensors and electronics is enabling the development of smaller, more agile UUVs, enhancing their suitability for specific niche applications, like inspection of underwater infrastructure or search and recovery operations.

The integration of unmanned systems into anti-submarine warfare (ASW) represents a significant paradigm shift, offering the potential to enhance detection, tracking, and ultimately, neutralization of submarines. Traditional ASW methods, reliant on manned aircraft, ships, and submarines, are expensive, labour-intensive, and vulnerable to enemy countermeasures. Unmanned surface vessels (USVs) and unmanned underwater vehicles (UUVs) offer a cost-effective and adaptable alternative, capable of persistent surveillance and extended deployments, while minimizing risk to human personnel.

USVs, acting as force multipliers, can enhance ASW capabilities by providing a persistent platform for deploying and recovering UUVs, relaying sensor data, and acting as a communications node. Equipped with advanced sensors, such as towed arrays or sonobuoys, USVs can detect acoustic signatures emitted by submarines. These sensors are deployed either directly from the USV or launched from it to cover larger areas and depths than the sensors onboard the USV alone. The deployment of multiple USVs, coordinated through a network, offers broad-area surveillance capabilities, greatly expanding the search area and increasing the likelihood of detecting even quiet submarines attempting stealthy maneuvers. Their ability to remain on station for extended periods, unlike manned ships constrained by crew fatigue and resupply requirements, significantly improves their effectiveness in ASW operations.

The limitations of USV-based ASW are primarily environmental. Oceanic conditions—high waves, strong currents, or dense fog—can hinder sensor performance and the deployment and retrieval of UUVs. Moreover, USVs operating on the surface are vulnerable to enemy detection and attack, necessitating the integration of defensive countermeasures such as electronic warfare systems and active decoys. However, these inherent risks are mitigated by the lower cost of losing a USV compared to a manned vessel.

UUVs represent the core of the unmanned ASW revolution. Operating submerged, UUVs are significantly more difficult to detect and track than surface vessels. Their ability to directly probe the underwater environment makes them uniquely suited to detecting and tracking submarines. Equipped with sophisticated sonar systems, they can passively listen for submarine acoustic signatures, or employ active sonar to actively search for submarines, mapping the underwater acoustic landscape in high detail. The use of passive sonar avoids the detection limitations of the active sonar approach. Passive acoustic arrays deployed from UUVs can detect and track even the quietest submarines, effectively extending the range and endurance of ASW operations significantly beyond the capabilities of traditional methods.

The types of sonar employed on UUVs vary greatly based on mission requirements. High-frequency sonar is ideal for close-range detection and target identification, while low-frequency sonar provides longer range detection, albeit with lower resolution. Synthetic aperture sonar (SAS) increases the resolution by using signal processing techniques. The choice of sonar depends on factors such as the operating depth, the target's acoustic signature, and the environmental conditions. UUVs may carry multiple sonar systems simultaneously, leveraging their different capabilities for a comprehensive understanding of the underwater acoustic environment.

Beyond sonar, UUVs can employ other sensors to enhance ASW capabilities. Magnetometers, for example, can detect the subtle magnetic anomalies generated by submarines. This sensor data complements that provided by sonar, offering an independent verification of submarine presence or potential locations. The fusion of these multiple sensor data streams through advanced algorithms allows for higher confidence and less ambiguity in identifying potential submarine targets. Furthermore, some advanced UUVs are equipped with optical cameras, which while limited in range by water turbidity, can provide valuable visual identification of targets when deployed in shallow, clear waters.

The autonomy of UUVs in ASW operations is a critical element. Modern UUVs are increasingly equipped with advanced AI capabilities allowing them to autonomously navigate complex underwater environments, evade obstacles, and make real-time decisions based on sensor data. This level of autonomy allows for extended patrols and enhanced mission effectiveness, enabling them to react to changing conditions and adapt to unexpected targets or threats. For example, an AI-powered UUV could autonomously adjust its search pattern based on the presence of detected acoustic anomalies, or prioritize tracking a perceived higher-threat target. This level of independent operation also dramatically reduces the operational workload on support vessels and shore-based personnel.

However, the deployment of UUVs in ASW operations also presents significant technological challenges. The immense pressures at greater ocean depths demand extremely robust and reliable underwater vehicle construction. The long endurance requirements necessitate energy-efficient propulsion systems and high-capacity energy storage solutions. Communication with submerged UUVs is another considerable hurdle. Acoustic communication, while feasible, is limited in bandwidth and range, necessitating efficient data compression techniques. The development of higher bandwidth and long-range underwater communication technologies remains an active area of research, focusing on approaches such as optical communication or advanced acoustic modems.

The cost of UUV acquisition and maintenance is substantial, though the potential for significantly increased operational efficiency and reduced risk to human personnel makes it a worthwhile investment in the long term. The potential for UUVs to revolutionize ASW is substantial but requires continued investment in research and development. The need to address challenges such as enhanced autonomy, improved communication technologies, and effective countermeasures is paramount. Successfully addressing these key technical hurdles will enable a new era in anti-submarine warfare, improving the capability of navies to detect, track, and counter submarine threats.

The operational integration of USVs and UUVs into existing ASW platforms and systems is also a key. Effective data fusion from multiple sensors and platforms is essential to create a complete picture of the underwater threat environment. This requires robust communication networks and sophisticated data processing algorithms capable of seamlessly integrating information from multiple sources. The development of effective command and control systems that can manage multiple USVs and UUVs simultaneously is also a critical requirement.

Furthermore, the integration of artificial intelligence (AI) and machine learning (ML) techniques are vital for improving the effectiveness of unmanned ASW systems. AI-powered algorithms can significantly improve automatic target recognition (ATR), enabling UUVs to autonomously identify submarines from other underwater objects. ML algorithms can analyze massive amounts of sensor data, aiding in predicting submarine movements and optimizing search strategies. The synergy between AI and sophisticated sensors is essential for optimizing the effectiveness of unmanned ASW platforms.

The future of ASW is undeniably tied to the advancement and integration of unmanned systems. The combined capabilities of USVs and UUVs offer significant advantages over traditional methods, providing cost-effective, persistent surveillance and the ability to operate in areas previously inaccessible to manned platforms. The further development of advanced sensors, autonomous navigation systems, enhanced communication technologies, and sophisticated AI capabilities will drastically increase the effectiveness of these unmanned platforms in detecting and countering submarine threats. This ongoing evolution will revolutionize ASW strategies and tactics in the years to come, ensuring maritime dominance and security in the face of evolving underwater threats. The silent, persistent vigilance of these underwater drones will define the future of anti-submarine warfare.

The preceding discussion centered on the defensive applications of unmanned naval systems, particularly in anti-submarine warfare. However, the potential of unmanned surface vessels (USVs) and unmanned underwater vehicles (UUVs) extends far beyond passive surveillance and reconnaissance. These platforms, with the incorporation of appropriate offensive payloads, represent a significant shift in the landscape of naval warfare, introducing the potential for autonomous drone-based naval attacks. This section will explore the feasibility, challenges, and implications of such a capability.

The most immediate application of offensively-equipped drones lies in the realm of coastal defence and asymmetric warfare. Smaller, less sophisticated USVs, equipped with relatively simple but effective payloads such as explosives or even shaped charges, could be deployed in swarms to overwhelm coastal defences or attack smaller, less heavily defended vessels. The low cost of these smaller drones mitigates the risk significantly; the loss of a single drone is far less consequential than the loss of a manned vessel. This strategy, employing swarm tactics, leverages sheer numbers to overcome enemy defences, potentially overwhelming radar and missile systems designed to counter individual, larger threats. The autonomous nature of these drones allows for coordinated attacks against multiple targets simultaneously, significantly increasing the effectiveness of the assault. Real-world examples of small drone swarms already seen in land-based conflicts provide a strong indication of the potential effectiveness of this approach in the maritime domain.

However, the application of swarm tactics against larger, more heavily defended vessels presents a greater challenge. While sheer numbers might overwhelm defences to some extent, the precision and lethality required to successfully engage larger ships, especially those with advanced anti-missile systems, necessitates more sophisticated payloads and targeting capabilities. The success of such attacks relies heavily on exploiting weaknesses in the target's defences, potentially focusing on less protected areas such as engine rooms or ammunition magazines. This demands a higher level of intelligence gathering and target analysis, relying heavily on sensor data fusion and advanced AI algorithms to identify the most vulnerable points for effective strikes.

Larger USVs offer a different approach. These platforms, possessing greater payload capacity and potentially more advanced sensor suites, could be equipped with heavier weaponry, such as anti-ship missiles or torpedoes. While this approach sacrifices some of the cost-effectiveness of smaller swarm tactics, it significantly increases the potential for successful engagement of larger vessels. The autonomous navigation capabilities of these larger USVs would allow them to navigate complex maritime environments, avoiding obstacles and enemy vessels, while the enhanced sensor systems would provide improved target acquisition and tracking. The integration of advanced AI would allow these drones to make real-time decisions based on the evolving tactical situation, adapting to enemy countermeasures and selecting the most effective course of action.

The use of UUVs in offensive operations presents unique possibilities. Their ability to operate submerged significantly reduces their vulnerability to detection and allows them to approach targets undetected. Equipped with appropriately sized torpedoes or other underwater explosives, these drones could launch attacks against vessels from beneath the surface. This approach represents a particularly significant threat to larger surface vessels, as the difficulty in detecting and countering submerged attacks presents a formidable challenge. The development of more effective and autonomous UUVs is essential to realizing this potential, however. The challenges associated with underwater communication and navigation at longer ranges remains significant obstacles to overcome.

However, several significant technological difficulties must be overcome before widespread adoption of drone-based naval attacks becomes reality. First and foremost is the development of highly reliable autonomous navigation systems capable of operating in complex and unpredictable maritime environments. This requires advanced AI algorithms capable of processing vast amounts of sensor data and making real-time decisions to avoid obstacles, navigate in adverse weather conditions, and effectively evade enemy detection. Advanced sensor fusion techniques will be critical to ensuring accurate target acquisition and tracking in the often-challenging visual and acoustic environments of the ocean.

The integration of effective offensive payloads also presents significant engineering challenges. The design of reliable and effective weapons specifically for unmanned platforms requires careful consideration of size, weight, power consumption, and safety. The development of smaller, more efficient explosives, improved guidance systems, and robust mechanisms for payload delivery will be essential. This requires a combination of advancements in materials science, propulsion systems, and ordnance design.

Effective countermeasures against drone attacks also represent a vital area of development. The increasing threat of these unmanned systems necessitates the development of sophisticated defence systems capable of detecting and neutralizing these platforms. This includes advanced radar and sonar systems, electronic warfare capabilities, and effective counter-drone technologies. The development of robust defence systems is a central element in maintaining a level playing field in naval warfare.

Moreover, the legal and ethical implications of employing autonomous weapon systems in naval attacks warrant careful consideration. International law regarding the use of force, the rules of engagement, and the accountability for actions undertaken by autonomous systems need to be critically addressed. The potential for unintended consequences and escalation of conflicts raises serious ethical concerns that must be thoroughly examined.

The integration of advanced AI and machine learning algorithms will be essential in achieving truly autonomous navigation for naval drones. Current systems rely heavily on pre-programmed routes and GPS data, rendering them vulnerable in environments where GPS is denied or disrupted. Future autonomous systems will need to seamlessly integrate diverse sensor data — sonar, radar, Lidar, visual images — to build a comprehensive understanding of their surroundings. This necessitates sophisticated algorithms capable of real-time data fusion, object recognition, and decision-making, allowing the drones to autonomously navigate complex, dynamic environments, including crowded waterways, challenging weather conditions, and even the detection and avoidance of underwater obstacles. This capability would dramatically increase the operational range and survivability of naval drones. Furthermore, advanced AI would enable adaptive route planning, adjusting the drone's course based on real-time threat assessments, and even utilizing deception tactics to evade detection.

Swarm capabilities represent another significant frontier in naval drone technology. The coordination and control of multiple drones operating as a cohesive unit will require robust communication networks and sophisticated algorithms for collective decision-making. While simple swarms might rely on pre-programmed behaviours, the true potential of swarms lies in their ability to adapt to dynamic situations and exhibit emergent behaviour. This means drones collectively react to changing circumstances, adapting their strategies in real time based on the actions of other drones and the enemy. For example, a swarm of USVs might autonomously distribute themselves to maximize target coverage, while adapting to countermeasures, adjusting their approach based on the effectiveness of each drone's engagement and the enemy's defensive actions. This requires the development of resilient communication systems resistant to jamming and interference and sophisticated AI algorithms capable of coordinating complex interactions amongst multiple drones in dynamic, potentially adversarial, environments. This distributed autonomy, where individual drones have some level of independent decision-making but still coordinate their actions effectively within the swarm, would dramatically enhance the survivability and effectiveness of a drone swarm. The ability for smaller drones to collectively accomplish tasks beyond their individual capabilities will lead to a significant increase in the overall operational capabilities of naval forces.

The development of swarm tactics also necessitates advanced algorithms for managing the allocation of tasks within a swarm. Not all drones need to perform the same functions; some might specialize in reconnaissance, while others engage in direct attacks or electronic warfare. The AI systems need to dynamically assign roles based on the tactical situation, ensuring optimal utilization of the swarm's overall capabilities.

The potential applications of autonomous navigation and swarm capabilities are far-reaching. In anti-submarine warfare (ASW), swarms of autonomous UUVs could cooperatively search vast areas of ocean, significantly enhancing the detection and tracking of submarines. Similarly, in mine countermeasures (MCM), autonomous USVs and UUVs could conduct coordinated sweeps of minefields, significantly reducing the risk to manned vessels. The combined use of autonomous USVs and UUVs offers a powerful combination; USVs could provide a broad area search, while UUVs could investigate suspected mine locations with greater precision. The synergy between these two types of unmanned systems allows for efficient and effective mine countermeasure operations, reducing the risks to human personnel and significantly increasing the efficiency of clearing operations.

In offensive operations, swarms of USVs could overwhelm enemy coastal defences, while UUVs could launch coordinated attacks against enemy ships from beneath the surface. This approach would exploit the strengths of both platforms, maximizing the impact while minimizing the risks to manned vessels. The combined use of USVs and UUVs allows for a multi-pronged approach; USVs could distract and engage enemy surface defences while the UUVs slip through undetected to target high-value assets. The combination of both autonomous surface and submerged platforms offers significant advantages over relying on a single type of unmanned system.

However, the realization of these capabilities faces significant challenges. The development of robust and secure communication systems for swarms is crucial, as it will become the central element around which the autonomy of the swarm is built. The potential for enemy jamming or disruption of these communications presents a significant threat, demanding secure, redundant communication protocols. The systems must be able to operate reliably in a heavily contested environment with the possibility of enemy signal jamming and disruption. Furthermore, the integration of advanced AI algorithms into these systems necessitates significant improvements in processing power and energy efficiency. The algorithms need to be computationally efficient, demanding minimal resources while providing effective decision-making in real-time. The development of advanced machine learning techniques to overcome the limitations of traditional algorithms will be critical.

However, implementing effective collaborative control requires careful consideration of several factors. First, it necessitates the development of robust algorithms capable of smoothly transitioning between autonomous and human control, ensuring seamless handoffs without disrupting the mission. These algorithms must also be able to handle unexpected situations, smoothly adapting to changing circumstances and handling potential failures or malfunctions in either the human or machine components.

Secondly, the development of trust between human operators and autonomous systems is essential for successful HMT. Operators must have confidence in the capabilities and reliability of the autonomous systems, knowing that they will perform as expected and make informed decisions. This trust is built through extensive testing and validation, demonstrating the system's capabilities and reliability in a variety of operational scenarios. Transparent communication about the system's decision-making processes is also essential for building trust and confidence.

Thirdly, addressing the ethical implications of shared control is vital. In situations where autonomous systems make critical decisions that have significant consequences, the question of accountability becomes critical. Establishing clear lines of responsibility and defining the roles and limitations of both human operators and autonomous systems is essential for ensuring ethical and responsible use of this technology.

The future of drone technology depends heavily on effectively integrating human expertise with machine intelligence. The development of intuitive HMIs, robust collaborative control algorithms, and a strong emphasis on building trust between humans and machines are crucial steps in realizing the full potential of drones.

The challenge lies in developing a system where each element complements the other, creating a potent combination that surpasses the sum of its individual parts. Addressing the technical, ethical, and legal complexities associated with HMT is fundamental for ensuring the safe, responsible, and beneficial implementation of this transformative technology. Only then can we fully unlock the potential of drones to address a wide range of challenges across military, civilian, and commercial sectors.

The future of drone technology is not merely a technological narrative; it is inextricably intertwined with geopolitical realities, presenting both unprecedented opportunities and significant challenges. The most immediate geopolitical impact stems from the proliferation of autonomous drone capabilities. As drones become increasingly sophisticated and readily available, the threshold for engaging in armed conflict could be significantly lowered. Non-state actors, previously constrained by the complexities of acquiring and deploying conventional weaponry, could gain access to autonomous drone systems, potentially destabilizing regions and exacerbating existing conflicts. This poses a significant challenge to international security, as the ability to launch precise, targeted strikes becomes democratized, potentially undermining the existing power structures and leading to a more chaotic global order. The ease of acquisition and deployment of relatively inexpensive, yet effective, autonomous drones could also empower smaller nations to project their power in ways previously unimaginable, reshaping the balance of power in specific regions and potentially sparking regional conflicts.

Furthermore, the development of advanced AI-driven swarm technologies presents a particularly disturbing prospect. Imagine coordinated swarms of hundreds or even thousands of drones, each equipped with lethal payloads, autonomously targeting infrastructure, military installations, or even civilian populations. The sheer scale and speed of such an attack would overwhelm traditional defence mechanisms, creating a new form of asymmetric warfare that challenges established notions of military strategy and defence capabilities. The potential for catastrophic damage and widespread casualties raises serious ethical and humanitarian concerns, necessitating the development of effective countermeasures and robust international regulations to prevent the misuse of such technology.

Beyond the military domain, the widespread adoption of civilian drones raises a overabundance of geopolitical concerns. The use of drones for surveillance, data collection, and potentially even targeted assassinations by government agencies raises critical questions about privacy, civil liberties, and the potential for abuse.

The potential for misuse of drone technology extends beyond military and civilian applications. The use of drones for smuggling, illegal trafficking, and other criminal activities is a serious concern. The relative ease with which drones can be modified for illicit purposes and the difficulty in tracking and intercepting them pose significant challenges to law enforcement agencies and border control authorities worldwide. This creates a need for international cooperation to develop effective countermeasures and establish international regulations to curb the misuse of drone technology for criminal activities.

The geopolitical implications of future drone capabilities necessitate a proactive approach towards international cooperation and regulation. Existing arms control agreements need to be revisited and updated to address the unique challenges posed by autonomous drone technology. The development of international norms and standards for the responsible development, deployment, and use of drones is central to prevent an escalation of conflict and mitigate the risks associated with their proliferation. This requires international dialogues and negotiations, involving both state and non-state actors, to establish a shared understanding of the ethical and security implications of drone technology and to develop effective regulatory frameworks.

The Geopolitical Implications of Drone Proliferation

The relatively low cost of entry for drone production and operation, compared to traditional military hardware, democratizes access to advanced military capabilities, potentially empowering non-state actors and smaller nations to challenge established regional powers. This shift in the balance of power introduces a significant degree of unpredictability into the international security landscape, with potentially cascading effects on regional and global stability.

One of the most significant drivers of regional instability stemming from drone proliferation is the inherent asymmetry it creates. A relatively small nation or even a well-funded non-state actor can, through the acquisition and deployment of drones, inflict significant damage on a much larger, more conventionally powerful adversary. This asymmetry undermines traditional deterrence mechanisms, as the cost-benefit calculation for aggression changes drastically. A conventional military power might hesitate to engage in a conflict where the potential for disproportionate losses from relatively inexpensive drone attacks is high, effectively lowering the threshold for conflict initiation by weaker actors. The 2020 Nagorno-Karabakh conflict serves as a stark example, where Azerbaijan's effective use of Turkish-supplied Bayraktar TB2 drones against Armenian forces significantly altered the course of the war, demonstrating the disruptive potential of drones in asymmetric warfare. This success story has spurred a wave of interest in acquiring similar drone capabilities, triggering a regional arms race in the South Caucasus and beyond.

The impact extends beyond direct military confrontation. The potential for drone attacks against civilian infrastructure, economic targets, or even civilian populations creates a climate of fear and uncertainty, exacerbating existing tensions and potentially driving humanitarian crises. This capability presents a potent tool for coercion and intimidation, allowing actors to project influence and destabilize their adversaries without engaging in full-scale conventional warfare. The use of drones for targeted assassinations and other extrajudicial killings raises significant ethical and legal concerns, further complicating the geopolitical landscape and potentially powering cycles of violence. The lack of clear legal frameworks governing the use of drones in armed conflict only amplifies these concerns.

The proliferation of drone technology also challenges established alliances and security architectures. The ability of smaller nations or non-state actors to acquire and deploy drones reduces their reliance on traditional allies for security provision, potentially weakening existing alliances and fostering a more multipolar and unpredictable security environment. This can lead to a strategic decoupling, as states may opt for a more independent approach to security, prioritizing their own acquisition of drone capabilities over traditional military cooperation. The potential for unexpected alliances and partnerships, driven by the desire to acquire or counter drone technologies, introduces a considerable element of risk and uncertainty in international relations.

The arms race driven by drone proliferation also stimulates a cycle of technological escalation. As one party develops more advanced drone capabilities, the other party is incentivized to respond in kind, leading to a constant escalation of offensive and defensive capabilities. This leads to significant investment in counter-drone technologies, a parallel arms race focused on disrupting and neutralizing drone threats. This constant innovation can lead to a technologically driven escalation, where advances in offensive drone capabilities quickly stimulate corresponding advances in defensive systems, in a feedback loop that may be difficult to manage. The development of increasingly sophisticated AI-driven countermeasures, autonomous defensive systems, and advanced electronic warfare capabilities adds another layer of complexity to this dynamic arms race.

Furthermore, the ease of acquiring drone technology introduces significant risks of unauthorized use and proliferation into the hands of terrorist organizations and other malicious actors. This poses a severe threat to global security, as these groups could leverage drone capabilities for devastating attacks against civilian targets or critical infrastructure, thereby significantly disrupting stability across regions. The ability to assemble and modify commercially available drones with improvised explosive devices (IEDs) lowers the technical barrier to entry, allowing even poorly equipped groups to engage in high-impact attacks. The relatively low cost and ease of concealment associated with drones also contribute to their appeal as weapons of choice for non-state actors.

The geographical spread of drone technology also exacerbates the challenges in regulating its use and preventing its misuse. Unlike traditional weapons systems, which often involve complex manufacturing processes and supply chains, drones can be produced in various locations, and the components can be sourced from diverse suppliers. This makes it significantly more difficult to enforce international arms control agreements and monitor the spread of this technology, leading to an increasing challenge in controlling the flow of potentially lethal autonomous systems into unstable regions.

The lack of transparency surrounding drone strikes, particularly those conducted by states outside their own territories, often leads to accusations of civilian casualties and violations of international humanitarian law. This erosion of trust between states and the undermining of international legal norms contribute to increased tensions and instability. The discussion around lethal autonomous weapons systems (LAWS), or "killer robots," also remains a highly contentious and ethically challenging issue, particularly regarding accountability for actions taken by autonomous drones.

The democratization of aerial warfare, a hallmark of drone proliferation, profoundly impacts asymmetric conflicts. No longer are air superiority and precision strikes the exclusive domain of major military powers. Non-state actors (NSAs), ranging from insurgent groups to terrorist organizations and even criminal cartels, are increasingly leveraging drone technology to challenge established states and reshape the dynamics of conflict. This shift presents a significant challenge to traditional counterterrorism and counterinsurgency strategies, demanding a reassessment of existing doctrines and the development of novel approaches.

The relative affordability and ease of acquisition of commercially available drones, coupled with readily available online tutorials and open-source modifications, dramatically lowers the barrier to entry for NSAs. While sophisticated military-grade drones remain beyond the reach of most NSAs, readily available quadcopters and smaller UAVs can be easily weaponized with improvised explosive devices (IEDs), providing a potent, if less precise, weapon capable of inflicting significant damage. This accessibility transforms the battlefield, allowing NSAs to conduct attacks with a level of sophistication previously unimaginable. The use of drones for reconnaissance, surveillance, and targeted attacks empowers NSAs to challenge superior forces with an asymmetric advantage.

The effectiveness of this approach is demonstrated across various conflict zones. In the ongoing conflicts in Iraq, Syria, Yemen, and other regions, various insurgent groups have employed drones for various purposes, including launching attacks against military bases, conducting targeted assassinations, and coordinating ground offensives. These actions demonstrate the increasing proficiency of NSAs in integrating drone technology into their operational strategies, showcasing their adaptability and willingness to exploit technological vulnerabilities.

The impact extends beyond direct military confrontation. The psychological impact of drone attacks on civilian populations and military personnel cannot be underestimated. The persistent threat of aerial surveillance and the potential for surprise attacks contribute to a climate of fear and uncertainty, disrupting daily life and potentially undermining the morale of both civilian and military targets. This constant psychological pressure is a potent form of asymmetric warfare, inflicting a cost that is difficult to quantify but nonetheless substantial. The ability of NSAs to leverage drones for propaganda and dissemination of information further amplifies their impact, creating a more comprehensive campaign of asymmetric warfare.

The strategic implications for counterterrorism and counterinsurgency are substantial. Traditional counterinsurgency strategies often rely on establishing physical dominance and control over territory. However, the mobility and relative invisibility of drones challenge this approach. The ability of NSAs to conduct attacks from afar and evade detection undermines traditional military superiority, demanding a recalibration of military responses.

The challenge lies in effectively countering drone threats without escalating conflicts. Traditional military responses, such as air strikes or ground assaults, may prove ineffective or even counterproductive, potentially leading to civilian casualties and stimulating anti-government sentiment. This necessitates the development of more sophisticated and nuanced counter-drone strategies. Investing in electronic warfare technologies, developing advanced jamming systems, and employing counter-drone weapons that can neutralize drones without causing widespread damage are critical aspects of this evolving counter-drone landscape. Furthermore, addressing the root causes of conflict and improving governance and security institutions within conflict-affected countries remain critical long-term strategies.

Furthermore, the proliferation of drone technology necessitates a renewed focus on international law and regulations. The use of drones in armed conflict raises complex legal and ethical questions, particularly concerning proportionality and distinction. The potential for civilian casualties caused by drone strikes, particularly by NSAs lacking clear lines of command and accountability, highlights the need for international guidelines and accountability mechanisms. The development of effective and enforceable international regulations is essential in mitigating the risks posed by the proliferation of drones and ensuring adherence to international humanitarian law.

The ethical considerations surrounding the use of drones by NSAs are equally pressing. The lack of transparency in the development, acquisition, and deployment of drones often obscures the chain of responsibility, potentially leading to impunity for human rights abuses. This necessitates an investigation into the various ways that NSAs acquire drone technology, the sources of funding and support, and the development of mechanisms for accountability when drones are used for unlawful purposes. This includes addressing the challenges posed by the international transfer of drone components and technologies.

The challenge of combating drone use by NSAs also extends to the realm of intelligence and information sharing. Developing robust intelligence capabilities to identify and track the movement of drones, their operators, and support networks is essential. Enhanced cooperation between states and international organizations is essential for effective intelligence gathering and sharing. This collaboration needs to encompass the sharing of technical information on drone technologies, methods of detection and countermeasures, and best practices for disrupting the supply chains that support the proliferation of weaponized drones.

The increasing sophistication of drone technology poses another challenge. As drone technology continues to evolve, NSAs will likely adapt and develop more advanced tactics, potentially integrating drones with other technologies to enhance their effectiveness. This requires a continuous effort to develop and adapt counter-drone strategies, ensuring that defence systems remain ahead of the curve. A proactive and adaptive approach, focusing on both technological advancements and the development of robust legal frameworks, is essential to mitigating the risks posed by drone proliferation in asymmetric warfare.

Ultimately, addressing the challenges posed by drone proliferation requires a multifaceted approach. This includes investing in advanced counter-drone technologies, strengthening international cooperation to regulate the transfer of drone technology, bolstering intelligence-gathering capabilities, and promoting sustainable solutions that address the underlying drivers of conflict and instability. The long-term security implications of drone proliferation cannot be underestimated, and a proactive approach is essential for maintaining global stability in an increasingly complex and uncertain geopolitical environment. The challenge is not merely technological; it necessitates a strategic reassessment of counterterrorism and counterinsurgency strategies, an evolution in international law, and a renewed focus on ethical considerations in the face of evolving warfare.

The proliferation of drone technology is not simply a phenomenon impacting asymmetric warfare; it is fundamentally reshaping the landscape of great power competition. Major powers are engaged in a technological arms race, developing and deploying increasingly sophisticated drones for a wide array of military and intelligence operations. This competition is characterized by a focus on enhancing capabilities, expanding operational ranges, and integrating advanced technologies like artificial intelligence and swarm capabilities. The strategic implications are profound, impacting the balance of power, altering military doctrines, and raising concerns about escalation and accidental conflict.

The geopolitical implications of this great power competition in drone technology extend far beyond military applications. The deployment of drones for intelligence gathering, cyber warfare, and even economic espionage is rapidly increasing. The potential for miscalculation and escalation is a serious concern. An accidental clash between drones operated by rival powers could easily spiral into a larger conflict. Moreover, the use of drones for cyber attacks and data theft presents a new dimension of information warfare, blurring the lines between traditional military conflict and other forms of state competition. This need the development of robust cyber security measures and strategies for mitigating the risks of these non-kinetic forms of conflict.

The ethical considerations surrounding the use of drones in great power competition are equally critical. The lack of transparency in the development and deployment of these advanced systems, particularly those incorporating AI, raises concerns about accountability and the potential for unintended consequences. The use of autonomous weapons systems, capable of making lethal decisions without human intervention, raises complex ethical questions about the nature of warfare and the potential for escalation. International discussions and agreements are essential for establishing norms and regulations for the responsible development and deployment of these technologies, ensuring compliance with international humanitarian law, and minimizing the risks of accidental escalation.

Furthermore, the arms race in drone technology is stimulating a parallel competition in counter-drone technologies. Major powers are investing heavily in developing sophisticated anti-drone systems, including electronic warfare capabilities, laser weapons, and artificial intelligence-powered detection and neutralization systems. This counter-drone technology is not only designed to defend against attacks, but also to neutralize or capture enemy drones for intelligence purposes. The rapid pace of innovation in both drone and anti-drone technology necessitates a continual assessment of the balance of power, and adjustments to military strategies and doctrines accordingly.

The economic implications of the great power competition in drones are also significant. The global drone market is booming, with major powers vying for market share, technological leadership, and strategic influence. The economic incentives driving drone development and proliferation, including the demand for both military and civilian applications, are also driving innovation. This creates a dynamic environment where the technological advancements in one sphere, such as civilian applications, can rapidly translate into military capabilities in a dual use. The competition is not just about securing military advantages; it also involves securing a position of leadership in this rapidly growing economic sector. This economic dimension further complicates the geopolitical landscape, as trade agreements and technological partnerships can influence the balance of power and create new strategic alliances.

The escalating proliferation of drone technology necessitates a parallel effort towards international cooperation and arms control. However, establishing effective regulations and preventing the unchecked spread of this potent technology presents significant challenges. The inherent difficulties stem from the dual-use nature of drones, the rapid pace of technological advancement, and the diverse geopolitical interests of various nations. Civilian drones, readily available commercially, can easily be adapted for military purposes, blurring the lines between legitimate uses and potentially armed.

This ambiguity complicates efforts to establish clear distinctions and effective controls.

One major difficulty is the lack of a universally accepted definition of what constitutes a "military drone." While some drones are clearly designed for combat roles, others, initially intended for civilian applications, can be modified to perform military functions. This grey area hinders the establishment of effective export controls and arms embargoes. Furthermore, the rapid pace of technological innovation makes any regulatory framework quickly outdated. New drone capabilities and functionalities emerge constantly, often outpacing the ability of international bodies to adapt their regulations. This continuous evolution requires agile and responsive regulatory mechanisms, capable of addressing emerging threats and technological advancements.

Existing international arms control treaties, designed largely for conventional weapons, often lack the specificity needed to address the unique challenges posed by drones. The complexity of drone systems, encompassing hardware, software, and related technologies, presents significant challenges to verification and compliance. Monitoring the production, transfer, and use of drones across national borders requires sophisticated technological capabilities and international cooperation, often lacking in the current international framework. The difficulty of tracing the origin of drone components, particularly in instances of illicit trafficking, further compounds the problem. The ease with which drones can be modified and upgraded also poses challenges to verification efforts. Existing treaties may focus on the final product, but overlook the components and the know-how to integrate and utilize these technologies effectively, creating loopholes and undermining the effectiveness of monitoring processes.

Geopolitical considerations further complicate efforts towards international cooperation. Major Powers, often the leading developers and exporters of drone technology, possess vested interests in maintaining their technological advantage and market share. This often translates into a reluctance to participate fully in international arms control initiatives that might limit their own capabilities or economic interests. Furthermore, the asymmetry in drone capabilities between developed and developing nations creates power imbalances that make the creation of fair and equitable regulations challenging. Developing countries, often lacking the resources and expertise to produce advanced drones domestically, may be less inclined to support restrictive regulations that could limit their access to this technology. This disparity presents a significant obstacle in reaching consensus on international norms and standards.

The lack of a cohesive international framework also presents opportunities for non-state actors. Terrorist groups and other armed non-state actors have readily adopted drone technology, using commercially available drones for surveillance, reconnaissance, and even attacks. The ease of acquisition and adaptability of commercial drones make them attractive tools for non-state actors, enabling them to conduct operations that were once unattainable without significantly greater resources. This increased accessibility necessitates international collaboration to curb the flow of technology that could be used for malicious purposes. This requires not only limitations on exports, but also collaboration on developing and implementing technical countermeasures and information sharing capabilities.

The development and deployment of autonomous weapons systems (AWS) present a particularly difficult challenge for international arms control. The prospect of lethal autonomous weapons making life-or-death decisions without human intervention raises fundamental ethical and legal concerns. The lack of human control raises concerns about accountability, potential for unintended escalation, and the adherence to international humanitarian law. The complexity of such systems, and the potential for unforeseen malfunction, adds to the challenges of establishing effective oversight and control. Negotiations surrounding LAWS are further complicated by differing national interests and priorities, resulting in protracted debate and lack of consensus on international regulation. Discussions surrounding a pre-emptive ban on these weapons have been underway, but the lack of agreement highlights the difficulties of reaching global consensus on emerging technologies.

Despite these challenges, there have been some positive developments in international cooperation on drone technology. Several international organizations, including the UN and the Organisation for Security and Co-operation in Europe (OSCE), have initiated discussions and established working groups focused on addressing the implications of drone proliferation. These initiatives, while still in their early stages, are important first steps towards building a more robust international regulatory framework. Such discussions emphasize the need for information sharing, capacity building, and the development of common standards for responsible drone development and operation. Collaboration on the development of counter-drone technologies also plays a role in mitigating risks and ensuring responsible use of this powerful technology.

Moreover, individual states have increasingly focused on enhancing their national regulations governing drone use, focusing on both civilian and military applications. These national regulations, while not always uniform, contribute towards establishing a more consistent approach toward this rapidly evolving technology. In the absence of a comprehensive global agreement, these national regulations provide a floor for the responsible development, testing, and deployment of drone technologies, at least within national limits. However, it's essential for states to align their domestic regulations with international norms and standards to enhance the effectiveness of global efforts to regulate the spread of drone technology.

In the longer term, the development of an effective international regime for drone governance will require a comprehensive approach that addresses the technological, geopolitical, and ethical challenges. This will likely involve a combination of international treaties, legally binding agreements, and cooperative initiatives focusing on capacity building and technology transfer. Furthermore, international cooperation on the development of technical countermeasures is essential to prevent the misuse of drone technology by non-state actors. The ongoing discussions within the UN and other international bodies represent a critical stage in this process. The success of these initiatives will hinge on the willingness of states to prioritize the development of a collaborative, transparent, and effective regulatory framework for managing this powerful technology. The alternative is a continued race toward an unstable future, with potentially devastating consequences. The need for urgent and concerted international action is undeniable.

The proliferation of drone technology, as witnessed in recent conflicts, fundamentally alters the scenery of global security. The ease of acquisition, relative affordability, and diverse functionalities of drones, ranging from reconnaissance to lethal strikes, empower both state and non-state actors in unprecedented ways. This democratization of military capabilities, while potentially beneficial in certain contexts, presents profound challenges to established power structures and raises serious concerns about regional stability and international order.

The future of global security in the age of widespread drone adoption will be characterized by a complex interplay of several key factors. First, the technological arms race will continue unabated.

States and non-state actors alike will constantly strive to develop more advanced drone systems, equipped with enhanced capabilities such as improved sensors, longer flight endurance, greater payload capacity, and more sophisticated autonomous functionalities. This relentless pursuit of technological superiority will fuel a cycle of counter-development, forcing nations to invest heavily in defensive technologies designed to counter the threat posed by drones. This will include the development of advanced anti-drone systems, ranging from electronic warfare technologies that jam or disrupt drone signals, to kinetic weapons designed to neutralize drones in flight. The development of sophisticated AI-powered systems to identify and track drones, and autonomously engage threatening drones, is inevitable. This will create new markets and dependencies, potentially exacerbating existing geopolitical tensions.

Secondly, the strategic implications of drone warfare will reshape geopolitical dynamics. Smaller states and non-state actors, previously reliant on asymmetrical warfare tactics, now possess the potential to project power and inflict significant damage on larger, more technologically advanced adversaries. This shift in the balance of power will necessitate a reassessment of traditional military doctrines and strategies. The vulnerability of critical infrastructure, military bases, and civilian populations to drone attacks need the development of new defensive strategies and capabilities. The potential for widespread drone swarms to overwhelm existing air defences highlights the need for innovative defensive measures. The future battlefield will likely witness an unprecedented level of drone-on-drone combat, forcing the development of sophisticated AI-driven countermeasures.

Furthermore, the ethical dilemmas surrounding the use of drones will become increasingly prominent. The ability to conduct targeted strikes remotely, with minimal risk to the operator, raises fundamental questions about accountability, proportionality, and the potential for civilian casualties. The lack of human presence on the battlefield also complicates the assessment of collateral damage and adherence to international humanitarian law. As autonomous capabilities of drones improve, the ethical concerns intensify. The potential for lethal autonomous weapons systems (LAWS) to make life-or-death decisions without human intervention necessitates a thorough examination of the legal and moral implications. The development of international norms and regulations overriding the use of drones, particularly LAWS, will be central in mitigating these risks and establishing a framework for responsible drone operations.

The proliferation of drones also poses significant challenges to international law and arms control. The dual-use nature of drone technology – the potential for civilian drones to be easily adapted for military purposes – complicates efforts to establish effective export controls and arms embargoes. Existing international treaties, designed for conventional weapons, often lack the specificity needed to address the unique challenges posed by drones. The rapid pace of technological advancement further complicates the development of effective regulatory mechanisms. The challenge is to create a legal framework that balances the legitimate uses of drone technology with the need to prevent its proliferation for malicious purposes, while addressing the unique challenges posed by autonomous weapons systems. This requires significant international cooperation and a willingness to adapt to the rapid pace of technological change.

The geopolitical implications extend beyond the battlefield. The economic implications of the drone industry are substantial. The growth of the drone market, encompassing both military and civilian applications, presents significant opportunities for economic growth and technological innovation. However, it also raises concerns about the potential for economic inequality, with technologically advanced states potentially widening their advantage over developing nations. Moreover, the security concerns associated with drone technology could hinder economic development and international trade.

The future of global security in the context of widespread drone adoption will also be profoundly affected by the interplay between state and non-state actors. Terrorist organizations and other armed non-state actors have already demonstrated their ability to acquire and utilize drones for various purposes, including surveillance, reconnaissance, and attacks. The ease of acquiring commercially available drones and the relative simplicity of modifying them for military use make them attractive tools for non-state actors seeking to undermine state authority. The threat posed by non-state actors utilizing drone technology necessitates a significant shift in counterterrorism strategies, requiring improved intelligence gathering capabilities, enhanced cyber security measures, and the development of effective counter-drone technologies. This also highlights the need for international cooperation to curtail the flow of technology to malicious groups and prevent the misuse of drones.

The development of effective countermeasures against drones will be critical in shaping the future of global security. These measures will likely involve a combination of technological, operational, and regulatory strategies. Regulatory changes will involve the implementation of stricter export controls, the establishment of international standards for responsible drone development and operation, and the development of legal frameworks to govern the use of autonomous weapons systems.

Looking ahead, the future of global security will depend heavily on the capacity of the international community to adapt to the realities of widespread drone proliferation. This requires a multifaceted approach: fostering international cooperation to establish effective arms control measures, promoting responsible development and use of drone technology, investing in robust countermeasures, and addressing the ethical implications of autonomous weapons systems. The challenge is to leverage the potential benefits of drone technology for civilian purposes while mitigating the significant risks to global security posed by its unchecked proliferation in the military domain. The continued absence of a comprehensive, globally accepted framework for regulating drone technology promises a future punctuated by both escalating conflict and increasing vulnerability.

Nano Drones

The burgeoning field of nano drone technology represents a significant leap forward in unmanned aerial vehicle (UAV) capabilities. Defining what constitutes a "nano drone" is central, as the term itself lacks a universally agreed-upon definition. However, a working definition can be established based on size, weight, and functionality. Generally, nano drones are characterized by their extremely small size, typically measuring less than 10 centimetres in length and weighing less than 100 grams. This minuscule scale distinguishes them sharply from their larger counterparts, presenting unique design challenges and opportunities. Unlike larger drones capable of carrying substantial payloads, nano drones are severely constrained by their limited carrying capacity. This limitation necessitates innovative approaches to both design and functionality.

The technological advancements that have propelled nano drones from theoretical concepts to tangible realities are remarkable. Miniaturization of crucial components is paramount. This includes the development of incredibly small, yet powerful, microprocessors, sensors, and actuators. Advances in micro electromechanical systems (MEMS) technology have been instrumental, allowing for the creation of highly integrated and compact systems. Significant progress in energy storage technology has also been essential. High-energy-density batteries, such as lithium-polymer batteries, are vital for providing sufficient flight time, even with the inherent limitations imposed by the nano drone's small size. The development of efficient power management systems is another key element, ensuring that limited energy resources are utilized optimally throughout the flight. Furthermore, breakthroughs in materials science have allowed for the creation of lightweight, yet durable, materials suited to the unique demands of nano drone construction. These materials must be strong enough to withstand the stresses of flight, yet light enough to minimize weight and maximize flight time.

The minuscule size of nano drones presents numerous challenges. One of the most significant is the limited payload capacity. This directly impacts the types of sensors and equipment that can be integrated. The small size also impacts control systems. Maintaining stable flight and accurate manoeuvrability in a small, lightweight frame is significantly more difficult than with larger drones. Environmental factors such as wind gusts can have a disproportionately larger effect on the stability of a nano drone. Furthermore, the limitations of miniaturization affect the range and endurance of the drone. The batteries needed to power even the simplest functionalities are still relatively bulky, leaving little room for additional equipment. This constraint often necessitates trade-offs between flight time, payload capacity, and sensor capabilities. The development of sophisticated algorithms for flight control and autonomous navigation is essential to mitigate these limitations.

Design considerations for nano drones are multifaceted and often involve compromises. Aerodynamics plays a central role in maximizing flight efficiency and minimizing energy consumption. The design must account for the effects of air resistance, lift generation, and stability at different speeds and altitudes. The choice of materials is critical, as it impacts the weight, durability, and manufacturing costs. Carbon fibber composites are commonly employed due to their high strength-to-weight ratio. However, other materials such as polymers and specialized alloys are also explored for specific applications. Durability is a key consideration, as nano drones may operate in harsh environments, requiring resistance to impacts, temperature extremes, and moisture. Furthermore, the design must account for the integration of various components, including the microprocessor, sensors, actuators, and power source, while keeping the overall size and weight as small as possible. These are highly constrained design considerations, demanding creative solutions and innovative engineering.

The construction of nano drones necessitates advanced manufacturing techniques. 3D printing, particularly with materials like polymers and composites, is proving increasingly valuable for rapid prototyping and customized production. However, the creation of incredibly small and intricate components often relies on advanced micro-assembly processes. This involves highly precise manipulation and placement of tiny components, requiring sophisticated robotic systems and meticulous quality control. The need for miniaturization extends to the manufacturing process itself, with precision engineering essential at every stage. While 3D printing offers flexibility and speed, the limitations of resolution and material properties may restrict its applicability for certain components. The integration of various manufacturing techniques, possibly combining additive manufacturing with traditional micro-fabrication methods, is likely to become commonplace to overcome these challenges. The scale of nano drone production is also a significant consideration. Mass production techniques will be vital for cost-effective deployments, especially in scenarios involving large swarms.

The small size of nano drones opens up new possibilities for deployment. Manual launch is possible in certain scenarios, particularly for single-unit deployments. However, for large-scale deployments involving swarms, aerial dispersal methods from larger aircraft become necessary. This involves releasing a large number of drones from an aircraft in flight, often employing specialized mechanisms for controlled release and deployment. Ground-based launchers could also be used, potentially incorporating automated systems for rapid deployment of multiple drones. Each method presents its own unique challenges and considerations. Aerial dispersal, for example, requires careful planning to ensure optimal drone distribution and minimize the risk of collisions. Ground-based launchers need to be robust enough to withstand repeated use and potentially hostile environments. Logistics and coordination are critical for large-scale deployments, requiring sophisticated command and control systems and effective communication protocols to ensure the coordinated operation of large drone swarms. The challenge extends beyond mere deployment; effective swarm management during and after deployment becomes vital for achieving mission objectives.

Beyond military applications, the potential uses of nano drones are extensive and impactful. In environmental monitoring, nano drones can be deployed for assessing air and water quality, monitoring wildlife populations, and detecting pollutants. Infrastructure inspection use nano drones to examine bridges, power lines, and pipelines for damage, offering a safer and more efficient approach than traditional methods. Search and rescue operations benefit from the agility and manoeuvrability of nano drones, enabling them to navigate difficult terrain and locate individuals in need. Precision agriculture is another promising field, with nano drones aiding in crop monitoring, targeted pesticide application, and optimized irrigation. Furthermore, delivery services could be revolutionized by the use of nano drones for carrying small packages over short distances. Each of these civilian applications demands further research and development, ensuring safe and effective integration into existing societal structures. Regulatory frameworks need to adapt to the unique capabilities and implications of nano drone technology, navigating the complex interplay between innovation and societal safety. The implications for future technology are immense.

Navigation and environmental awareness are vital for any autonomous vehicle, and nano drones are no exception. Miniature

Global Positioning System (GPS) receivers provide location data, allowing the drone to pinpoint its position relative to a known coordinate system. However, the limitations of size necessitate the use of extremely low-power GPS chips, which may result in lower accuracy or slower acquisition times compared to larger counterparts. To compensate for GPS limitations and provide inertial data even during GPS outages, nano drones incorporate Inertial Measurement Units (IMUs). These IMUs typically consist of accelerometers and gyroscopes, measuring linear acceleration and angular velocity respectively. The data from the IMU, combined with GPS data (when available), allows the drone to estimate its position, velocity, and orientation, even in challenging environments or during periods of GPS signal loss. Sophisticated algorithms are employed to fuse the data from these different sensors, filtering out noise and improving overall accuracy.

Beyond basic navigation, nano drones require a wide array of sensors to perceive their surroundings and interact with their environment. Cameras are a critical component, offering visual information for navigation, target identification, and environmental assessment. The miniaturization of camera technology has been remarkable, with the development of extremely compact, high-resolution cameras that consume minimal power. These cameras often utilize CMOS (complementary metal-oxide-semiconductor) image sensors, chosen for their low power consumption and relatively high performance in small form factors. Different types of cameras may be employed depending on the specific application. High-resolution cameras are ideal for detailed images, while wide-angle lenses provide a broader field of view, useful for navigation and situational awareness. Thermal cameras, sensitive to infrared radiation, provide imaging capabilities independent of visible light, making them useful for surveillance in low-light conditions or detecting heat signatures. Beyond visual sensors, nano drones can incorporate other types of sensors depending on their mission profile. These might include proximity sensors to detect nearby obstacles, pressure sensors to measure altitude, magnetic sensors (magnetometers) for compass readings, and even chemical sensors for detecting specific substances in the air or water.

The power source is a critical constraint for nano drones, directly impacting their flight time and operational range. While significant advancements have been made in miniaturizing battery technology, limitations persist. Lithium-polymer (LiPo) batteries are currently the dominant choice, offering a high energy density relative to their weight and size. However, these batteries have inherent limitations in terms of energy density, cycle life, and safety, demanding careful consideration in the design and operation of nano drones. Research continues into improving LiPo battery technology, aiming for higher energy density and improved safety characteristics. Solid-state batteries are a promising area of investigation, potentially offering significant advantages over LiPo batteries in terms of safety and lifespan.

Beyond traditional battery technology, significant efforts are underway to explore alternative power sources for nano drones. Energy harvesting, the process of capturing and converting ambient energy into usable electrical power, offers the potential to significantly extend flight time. One promising avenue is solar energy harvesting. The integration of miniaturized solar cells onto the drone's body can provide supplementary power, extending the operational time, particularly in sunny conditions. The challenge lies in designing efficient and lightweight solar cells capable of generating sufficient power while minimally impacting the overall weight and aerodynamics of the drone. Another potential approach is vibrational energy harvesting, converting mechanical vibrations into electrical energy. This technique could prove particularly useful in environments with significant vibrations or movement, converting ambient energy into usable power for the drone's systems. The efficient integration of these alternative power sources requires innovative design and careful consideration of the trade-offs between weight, power output, and overall system efficiency.

At the core of any nano drone's intelligence lays the microprocessor and the associated software. Microprocessors used in nano drones are highly specialized, prioritizing low power consumption and high computational efficiency within a small form factor. These processors are typically based on ARM architectures, chosen for their low power consumption and extensive software support. The small size of these processors necessitates highly integrated designs, incorporating significant functionality onto a single chip. The software running on these processors plays a crucial role in autonomous operation, managing various onboard systems, and executing sophisticated algorithms for navigation, sensor fusion, and decision-making.

Autonomous operation requires sophisticated algorithms for flight control, obstacle avoidance, and navigation. These algorithms typically utilize sensor data from various onboard sensors, implementing advanced control systems to maintain stable flight and precise manoeuvrability. Obstacle avoidance algorithms leverage data from proximity sensors and cameras to detect obstacles and autonomously manoeuvre around them. Path planning algorithms optimize flight paths to reach target destinations while avoiding obstacles and maximizing energy efficiency. These algorithms are often implemented using advanced techniques such as artificial neural networks (ANNs) or machine learning (ML), enabling adaptive behaviour and improved performance over time.

The AI component of nano drones is essential for enabling complex functionalities, including autonomous decision-making. AI algorithms can be implemented to enable target recognition, identification, and tracking. This involves sophisticated image processing techniques, pattern recognition, and machine learning algorithms to analyze visual data from cameras and identify specific objects or individuals. These algorithms can be trained on large datasets of images and videos to improve accuracy and robustness. The use of AI also extends to mission planning, enabling nano drones to autonomously adapt to changing environmental conditions and execute complex tasks with minimal human intervention.

The miniaturization of these components is a testament to ongoing advancements in micro-electromechanical systems (MEMS) technology, allowing the creation of incredibly small and powerful devices with low power consumption. The design and manufacturing of nano drones demand cutting-edge techniques, often involving the integration of multiple materials and fabrication processes. Precision engineering is paramount, ensuring the flawless integration of all components within the highly constrained space available. Further research and development efforts focus on enhancing the capabilities of these components, including improved sensor resolution, longer battery life, higher computational power, and more robust AI algorithms.

Military Applications of Nano Drones

The transition from civilian applications to the military domain reveals a dramatic shift in the strategic implications of nano drone technology. While environmental monitoring and infrastructure inspection highlight the beneficial aspects of these miniature unmanned aerial vehicles (UAVs), their military applications introduce a new dimension of complexity, demanding careful consideration of ethical, operational, and strategic ramifications. In the realm of surveillance and reconnaissance, nano drones offer unparalleled capabilities, transforming traditional intelligence gathering methods.

Their diminutive size provides an inherent advantage in covert operations. Unlike larger drones, nano drones can easily infiltrate heavily guarded areas, traverse complex urban landscapes, or navigate dense forests undetected. Their ability to operate silently, powered by miniature, efficient motors, further enhances their stealth capabilities, making them ideal for clandestine surveillance missions. Equipped with high-resolution cameras, infrared sensors, and advanced microphones, these drones can gather a wealth of visual, thermal, and acoustic intelligence. This multi-sensor approach creates a more complete picture of the operational environment, providing critical situational awareness for military commanders.

The integration of advanced artificial intelligence (AI) algorithms significantly amplifies the effectiveness of nano drone surveillance and reconnaissance. AI-powered image recognition can automatically identify targets of interest, such as specific individuals, vehicles, or weapons systems, from the vast amount of data collected. This automated target recognition (ATR) reduces the workload on human analysts, allowing for faster processing of intelligence and quicker response times. Moreover, AI can enhance the drones' autonomous navigation capabilities, allowing them to navigate complex terrains, avoid obstacles, and adapt to changing environmental conditions with minimal human intervention. Machine learning algorithms can also be used to improve the accuracy of target identification over time, refining the drones' performance based on accumulated data.

Consider the potential application in urban warfare. A swarm of nano drones, each equipped with a miniature camera and microphone, could be deployed to gather intelligence within a densely populated city. These drones could autonomously navigate through the urban canyons, avoiding detection while transmitting real-time video feeds and audio recordings to a central command post. This information would provide invaluable situational awareness to ground troops, allowing them to anticipate enemy movements and plan their operations accordingly. Furthermore, the use of AI-powered facial recognition could allow for the identification of high-value targets within the urban environment. These capabilities greatly improve the precision and effectiveness of military operations, reducing collateral damage and minimizing risks to friendly forces.

Similarly, nano drones excel in battlefield reconnaissance. Their small size and manoeuvrability enable them to penetrate enemy territory with relative ease, collecting crucial intelligence on enemy positions, troop deployments, and weapon systems. Equipped with specialized sensors, they can detect the presence of enemy vehicles, artillery pieces, or other military assets. Their ability to operate in challenging environments, such as mountainous terrain or dense jungles, significantly expands their operational scope. The use of multiple drones in coordinated swarms further enhances their effectiveness, providing broad coverage and a comprehensive picture of the battlefield. The data collected by these drones can be used to inform strategic decision-making, optimizing military tactics and enhancing the effectiveness of combat operations. This information can drastically improve the effectiveness of military operations by enabling commanders to make well-informed decisions, minimizing risks and maximizing tactical advantages.

Beyond simple visual and acoustic data, nano drones can be equipped with a variety of advanced sensors to expand their intelligence gathering capabilities. For instance, chemical sensors can detect the presence of explosives or biological weapons, while radiation sensors can identify nuclear materials. These capabilities are central in identifying potential threats and ensuring the safety of military personnel. The data collected by these sensors can provide early warnings of impending attacks, allowing for timely countermeasures. The combination of these different sensor types within a single nano drone platform, or a coordinated swarm of drones, allows for comprehensive and synergistic intelligence gathering, surpassing the capabilities of any single sensor system.

However, the deployment of nano drones in military contexts also presents significant limitations. One critical concern is their vulnerability to detection. Advanced radar systems, infrared sensors, and acoustic detection technologies can be used to detect and neutralize nano drones, potentially rendering them ineffective. Countermeasures, such as advanced stealth technologies and jamming systems, are decisive in addressing this vulnerability. The development of counter-countermeasures, constantly evolving in an arms race, is vital to maintain operational effectiveness. The miniaturization of the drones necessitates equally miniaturized countermeasures, pushing the technological boundaries of both offense and defense.

Another challenge is the susceptibility of nano drones to environmental factors. Adverse weather conditions, such as strong winds or heavy rain, can significantly impact their flight performance and operational effectiveness. This limitation underscores the need for robust drone designs capable of withstanding challenging environmental conditions. Furthermore, the limited endurance of nano drone batteries restricts their operational range and mission duration, requiring careful mission planning and potentially necessitating the use of multiple drones in relay formations to extend their coverage.

The technological advancements enabling such missions are considerable. Miniaturized explosives, guided by sophisticated AI algorithms and GPS systems, allow for pinpoint accuracy in targeting specific individuals or assets. Nano drones equipped with high-explosive warheads, while carrying a relatively small payload compared to conventional munitions, can be devastatingly effective when deployed in swarms. The swarm tactic, leveraging the sheer number of individual drones, presents a formidable challenge to traditional defensive systems. A coordinated swarm can overwhelm defenses by saturating them with multiple, simultaneous attacks, rendering point-defense systems ineffective. Each drone can attack from a different vector, exploiting weaknesses and creating a highly effective, distributed assault. This presents a new paradigm for warfare, requiring a fundamental rethink of defensive strategies.

However, the very capabilities that make nano drone swarms effective also introduce significant limitations and risks. The small size of the explosive payloads, while beneficial for precision strikes, also limits their destructive power. Larger, heavily fortified targets may require multiple drone strikes, increasing the risk of collateral damage and exposing the swarm to countermeasures. Furthermore, the reliance on GPS and other electronic guidance systems makes the drones vulnerable to jamming or spoofing. Disrupting the navigation systems of even a small fraction of a swarm can significantly reduce its effectiveness, potentially turning a coordinated attack into a chaotic and ineffectual barrage. The risk of friendly fire incidents, especially in complex or contested environments, cannot be ignored.

Maintaining effective human control over nano drone swarms in combat situations presents a significant technological and operational challenge. The sheer number of drones involved necessitates the development of sophisticated command and control systems capable of managing a multitude of individual units simultaneously. Real-time decision-making, particularly in dynamic environments, requires advanced AI algorithms capable of interpreting sensory data and making autonomous adjustments to the attack plan. The potential for system failures or unintended consequences is substantial, demanding robust failsafe mechanisms and clear protocols for human intervention. The challenge lies in balancing autonomy and human oversight, ensuring that the drones remain effective instruments of warfare while preventing uncontrolled escalation or unintended harm.

The ethical and legal implications of using nano drones in targeted strikes are profound. The question of accountability in case of civilian casualties or unintended harm remains a crucial unresolved issue. Existing international humanitarian law struggles to adapt to the unique challenges presented by autonomous weapons systems. The potential for misidentification or targeting errors, exacerbated by the limitations of AI and the inherent difficulties of operating in complex environments, demands careful consideration. Furthermore, the potential for misuse or proliferation of nano drone technology cannot be overlooked. The ease of production and deployment, coupled with their relatively low cost, raises concerns about their availability to non-state actors or hostile regimes, potentially destabilizing global security.

The development of effective countermeasures is crucial in mitigating the threats posed by nano drone swarms. Advanced radar and sensor technologies, capable of detecting and tracking the small, fast-moving drones, are under development. Directed energy weapons, such as lasers and high-powered microwaves, offer the potential to neutralize nano drones in flight. The use of drone nets, acoustic sensors and electronic jamming systems are also being explored as effective countermeasures. However, the ongoing technological arms race between offensive and defensive capabilities means that the development of countermeasures must continually adapt to the evolving capabilities of nano drones. The challenge lies in developing systems that are effective, scalable, and cost-effective, capable of neutralizing large-scale swarm attacks while minimizing collateral damage and civilian casualties.

Beyond the technological and ethical considerations, the operational realities of deploying nano drones in combat scenarios must be carefully examined. Logistics and supply chains for deploying, maintaining, and resupplying large quantities of nano drones pose significant logistical challenges. The need for reliable communication systems and robust data processing infrastructure underscores the reliance on sophisticated support systems. The vulnerability of these systems to cyber attacks and electronic warfare also represents a significant security risk. Therefore, the successful integration of nano drones into military operations requires a comprehensive understanding of the entire operational ecosystem, addressing not only the technological capabilities of the drones themselves but also the supporting infrastructure and security measures required for effective and responsible deployment.

The prospect of employing nano drones in a coordinated swarm represents a paradigm shift in military strategy and tactics. No longer constrained by the limitations of individual units, the combined capabilities of a large-scale swarm offer a potential for overwhelming enemy defences and executing complex, synchronized attacks. This section delves into the intricacies of swarm behaviour, exploring the advantages, challenges, and vulnerabilities associated with this emerging battlefield paradigm.

One of the most significant advantages of swarm tactics lies in their ability to overwhelm conventional defensive systems. A single nano drone, with its limited payload and size, might be easily neutralized. However, a coordinated swarm, comprising hundreds or even thousands of individual units, presents a vastly different challenge. Each drone, operating independently yet in concert with its neighbours, can attack from a unique vector, saturating defensive systems and exploiting vulnerabilities. Imagine, for example, a swarm targeting an air defence system. Instead of a single, predictable attack, the swarm can approach from multiple directions, altitudes, and velocities, making it incredibly difficult for the system to track and engage all incoming threats simultaneously. This distributed attack significantly increases the likelihood of at least some drones successfully reaching their target.

The effectiveness of a swarm relies heavily on sophisticated algorithms governing their cooperative behaviour. These algorithms, typically powered by advanced AI, must manage communication, coordination, and decision-making among the individual drones. Several prominent algorithmic approaches are currently being explored. One approach involves a decentralized control system, where each drone operates autonomously based on local information and simple rules. These rules might dictate, for example, maintaining a certain distance from other drones, avoiding obstacles, or targeting specific points of interest. Through emergent behaviour, the collective action of individual drones can lead to surprisingly complex and effective coordinated movements. This decentralized approach is inherently robust, as the failure of one or even several drones does not necessarily compromise the entire swarm's mission.

Another approach to swarm control involves a more centralized system, where a central command unit dictates the actions of the individual drones. While this offers greater precision and control, it also makes the swarm more vulnerable to disruption. If the central command unit is compromised – perhaps through electronic warfare or cyber attack – the entire swarm can be rendered ineffective. Therefore, a hybrid approach, combining elements of both decentralized and centralized control, is likely to prove most effective in real-world scenarios. This hybrid approach allows for flexibility and adaptability while mitigating the vulnerabilities associated with each individual approach.

The technological challenges associated with effective swarm coordination are considerable. Real-time communication and data sharing among numerous drones require high-bandwidth, low-latency communication networks. The inherent limitations of miniaturized onboard systems, power constraints, and the need for robust communication protocols necessitate significant advancements in both hardware and software technologies. Furthermore, the algorithms governing swarm behaviour must be capable of adapting to dynamic and unpredictable environments. The swarm needs to be able to respond to unexpected obstacles, enemy countermeasures, and changes in the mission objectives. This requires sophisticated AI algorithms capable of interpreting sensory data, predicting enemy actions, and making autonomous adjustments to the swarm's behaviour.

However, the very characteristics that make nano drone swarms so effective also expose them to certain vulnerabilities. Their reliance on communication networks makes them susceptible to electronic jamming and disruption. A coordinated effort to jam or spoof the communication signals can disrupt the swarm's coordination, rendering it ineffective. Similarly, directed energy weapons, such as lasers or high-powered microwaves, can be particularly effective against the small, relatively fragile nano drones. These weapons can target individual drones or even disrupt the entire swarm by overwhelming them with concentrated energy bursts.

The development of effective countermeasures is an ongoing arms race. Research into advanced radar systems capable of detecting and tracking nano drones is actively underway. These systems must be able to differentiate between nano drones and other airborne objects, such as birds or insects, to minimize false positives. Acoustic sensors, which detect the sound signatures of drone propellers, are also being explored. Furthermore, drone nets and other physical barriers can be deployed to physically capture or incapacitate nano drones. The challenge lies in developing scalable and effective countermeasures that can neutralize large-scale swarm attacks without causing excessive collateral damage. The development of AI-powered countermeasures capable of predicting and intercepting swarm attacks is also a key area of research.

Beyond the technological aspects, the operational realities of employing nano drone swarms require careful consideration. The logistics of deploying, maintaining, and resupplying large numbers of drones pose considerable logistical challenges. The drones need a robust command and control system, reliable power sources, and effective communication networks. The potential for cyber attacks and electronic warfare against these systems necessitates robust cyber security measures. Furthermore, the ethical implications of employing autonomous weapons, even at the nano scale, cannot be ignored. Questions of accountability, proportionality, and the potential for civilian casualties demand careful consideration and the development of stringent operational protocols.

One of the most promising avenues of integration lies in the collaboration between nano drones and larger unmanned aerial vehicles (UAVs). Imagine a scenario where a larger UAV, acting as a mother ship, deploys a swarm of nano drones upon reaching a target area. The UAV, with its superior range, endurance, and payload capacity, can carry a substantial number of nano drones and deliver them precisely to the designated location. This deployment strategy bypasses many of the inherent limitations associated with launching nano drones individually, particularly in challenging or contested environments. Upon deployment, the nano drones could fan out to conduct reconnaissance, surveillance, or even targeted attacks, while the UAV maintains a safe distance, providing communication relay and overall situational awareness. This synergistic relationship, combining the long-range capabilities of the UAV with the stealth and manoeuvrability of the nano drones, represents a powerful combination.

Furthermore, the UAV could serve as a crucial data hub, processing information gathered by the nano drones and transmitting it back to command centres. Nano drones, due to their size and power constraints, possess limited data processing and communication capabilities. The UAV can alleviate these limitations by acting as a relay station, aggregating the data from multiple drones, processing it, and transmitting the most critical information to command and control systems. This function significantly enhances the speed and efficiency of situational awareness, providing commanders with real-time insights into the operational environment. This collaborative approach optimizes resource allocation, reducing communication bottlenecks and ensuring the seamless flow of data throughout the operational theatre. The potential for enhanced real-time intelligence gathering and rapid decision-making is immense.

Beyond UAVs, the integration of nano drones with manned aircraft offers equally compelling advantages. A fighter jet, for example, could deploy a swarm of nano drones to scout ahead, identify enemy targets, or even disrupt enemy defences before initiating an attack. The nano drones could provide real-time reconnaissance, pinpoint the location of high-value targets, and assess the effectiveness of attacks, all while minimizing the risk to the manned aircraft. This integration provides an added layer of tactical flexibility, enabling the pilot to make informed decisions based on real-time information gathered by the nano drones. This approach extends the reach and capabilities of manned aircraft significantly, enhancing their effectiveness in combat operations.

Integration with ground-based command and control systems is equally vital for realizing the full potential of nano drones. These systems must be able to effectively manage and coordinate the actions of large numbers of drones in a dynamic environment. This requires robust communication networks, sophisticated algorithms for swarm control, and advanced interfaces that allow operators to easily monitor and direct drone activity. The system needs to handle real-time data feeds from multiple drones, provide visual representations of the operational environment, and allow operators to adjust drone behavior in response to changing circumstances. This requires seamless data fusion and integration with other sensor networks, transforming raw data into actionable intelligence.

Data analysis platforms play a crucial role in harnessing the immense amount of data generated by nano drone swarms. These platforms must be able to efficiently process, analyze, and visualize the information gathered by the drones, extracting critical intelligence and providing timely insights to decision-makers. Advanced algorithms for image processing, pattern recognition, and predictive modelling can be employed to automate data analysis, identifying patterns and anomalies that might otherwise be missed. The ability to quickly identify and track specific individuals, vehicles, or other objects within a large swarm of nano drones is essential for effective targeting and operational efficiency. The creation of sophisticated visualizations that allow operators to intuitively understand the data and make informed decisions is crucial.

The integration of nano drones with other military systems is not without its challenges. One significant hurdle lies in ensuring seamless interoperability between disparate systems. Different systems often use different communication protocols, data formats, and operating systems, which can hinder the free flow of information. Developing standardized interfaces and protocols is critical to facilitate seamless data exchange and coordinated actions. This standardization is essential to improve system effectiveness and prevent communication bottlenecks that could compromise mission success. The complexity of integrating multiple systems, each with its own unique specifications and capabilities, requires careful planning and coordination.

Moreover, maintaining the security of the integrated system is crucial. The vulnerability of communication networks to cyber attacks and electronic warfare necessitates robust cyber security measures. The protection of sensitive data and the prevention of unauthorized access are paramount to preventing mission compromise. This requires the implementation of sophisticated encryption techniques, intrusion detection systems, and secure communication protocols. The development of resilient communication systems that can withstand cyber attacks and electronic warfare is essential for the long-term success of nano drone deployment.

Another area that demands attention is the development of effective human-machine interfaces (HMIs). The sheer volume of data generated by nano drone swarms can easily overwhelm human operators. Intuitive and user-friendly HMIs are essential to prevent operator overload and ensure timely decision-making. This requires the development of sophisticated visualization tools and advanced algorithms that can automate data analysis and alert operators to critical events. Efficient and intuitive interfaces that minimize cognitive workload and facilitate rapid response to dynamic situations are critical for mission success. These interfaces must be designed with the operational realities of the battlefield in mind, ensuring simplicity and effectiveness under pressure.

One plausible, albeit currently hypothetical, scenario involves utilizing nano drone swarms for urban reconnaissance and surveillance. Imagine a high-stakes hostage situation within a densely populated city. Deploying a large swarm of nano drones, each equipped with advanced sensors and miniature cameras, could provide unprecedented situational awareness without the risks associated with deploying larger, more detectable UAVs or manned aircraft. These nano drones could infiltrate buildings, navigate complex urban environments, and relay real-time video and other sensory data to command centers. This data could pinpoint hostage locations, identify the perpetrators, and monitor their movements, significantly enhancing the effectiveness of rescue operations and potentially minimizing casualties. The small size and stealth capabilities of nano drones allow for covert operations, minimizing the risk of detection and providing tactical advantages that larger drones cannot match.

The challenges in this scenario are significant, however. The sheer volume of data generated by a large swarm would require highly sophisticated data processing and analysis capabilities to filter out irrelevant information and highlight critical details in real-time. Moreover, maintaining communication and control over a large number of nano drones in a congested urban environment poses a significant technological hurdle. Signal interference, environmental obstructions, and the potential for drone malfunctions all need to be carefully addressed. The development of robust communication protocols and highly reliable swarm control algorithms are crucial to overcoming these challenges. Furthermore, the ethical implications of such widespread surveillance must be considered, ensuring compliance with international law and respecting individual privacy rights.

A second hypothetical scenario concerns the use of nano drones for detecting and neutralizing improvised explosive devices (IEDs). Nano drones, equipped with advanced sensors capable of detecting chemical signatures, could be deployed to scan potentially hazardous areas for IEDs. Their small size would allow them to access areas inaccessible to larger robots or human personnel, significantly reducing the risk to personnel involved in explosive ordnance disposal (EOD). If an IED is detected, the nano drone could then relay its location and provide high-resolution images to EOD technicians, enabling them to plan and execute a safe disposal procedure. This capability is especially valuable in environments where IEDs are commonly employed, such as conflict zones or areas of insurgent activity.

However, the deployment of nano drones for IED detection also presents challenges. The sensitivity of the sensors must be carefully calibrated to avoid false positives, which could lead to wasted time and resources. Robust communication is critical to ensuring the timely transmission of vital information to EOD teams. Furthermore, the potential for nano drones to be jammed or disabled by adversaries necessitates the development of countermeasures, such as redundant communication channels and enhanced resilience against electronic warfare. Finally, the use of nano drones in this context raises ethical questions regarding the potential for misuse or unintended consequences, requiring careful consideration of operational protocols and safety measures.

Consider now the potential application of nano drones in maritime environments. A fleet of nano drones, dispersed across a large area of ocean, could serve as a distributed sensor network, monitoring shipping lanes, detecting submarines, or tracking the movements of vessels of interest. Their small size and low power requirements allow for long-duration deployments with minimal logistical support. Their low-cost would make them ideal for deploying large numbers, providing comprehensive surveillance coverage.

The practical challenges of using nano drones in this context include the harsh maritime environment. Saltwater corrosion, high winds, and wave action can severely impact their functionality and longevity. Moreover, the need for effective communication over vast distances and in challenging environments is critical. Development of robust, waterproof, and highly resilient drones is crucial. Ensuring sufficient battery life for extended missions is another key constraint. The integration of efficient solar-powered energy harvesting systems would enhance the endurance of such deployments. Furthermore, the potential for loss of nano drones due to accidents or enemy actions needs to be considered within a broader operational strategy.

Moving beyond hypothetical scenarios, we can examine elements of existing drone technology that foreshadow nano drone capabilities. Current military drone programs, like the use of small, quadcopter-type UAVs for reconnaissance and surveillance, provide a glimpse into the potential of their miniature counterparts. The ability of these drones to operate autonomously, albeit with limitations in size and endurance, is a critical stepping stone in the development of nano drone technology. The successful employment of these smaller drones in various military contexts provides a foundation upon which to build for future nano drone applications. Analyzing the successes and failures of these existing systems will be crucial in shaping the development and deployment of nano drones.

While the majority of current military drone deployments involve larger, more capable systems, the trends towards miniaturization and increased autonomy strongly suggest that nano drones represent a logical next step in the evolution of this technology. The challenges in terms of power, communication, and control remain substantial, but ongoing research and development efforts are continually pushing the boundaries of what is possible. The integration of advanced AI algorithms and machine learning will play a critical role in overcoming these challenges, paving the way for fully autonomous nano drone swarms capable of operating effectively in diverse and demanding environments.

The analysis of real-world examples, although currently limited due to the nascent stage of nano drone technology, should also consider the development of countermeasures. The same advancements driving nano drone technology also spur the creation of systems designed to detect, disable, and neutralize these miniature threats. This arms race will shape the future of nano drone deployment, emphasizing the need for robust countermeasures and adaptive strategies to maintain a tactical advantage. As nano drone technology matures, so too will the countermeasures employed against them, leading to a continuous cycle of innovation and adaptation on both sides.

Countermeasures Against Nano Drones

The effective countermeasure against any threat begins with accurate detection and precise tracking. This is particularly challenging with nano drones due to their diminutive size, potential for swarming behaviour, and inherent ability to exploit gaps in existing surveillance systems. Successfully countering these threats requires a multi-layered approach leveraging diverse detection technologies and sophisticated data fusion techniques.

One of the most established technologies for object detection, regardless of size, is radar. However, traditional radar systems, designed to detect larger aircraft or missiles, may struggle with nano drones due to their small radar cross-section. The weak signals reflected by these tiny devices can easily be lost within background noise or masked by clutter in dense urban or natural environments. To address this limitation, researchers are exploring the use of advanced radar technologies, including high-frequency radar systems and advanced signal processing algorithms. High-frequency radar offers improved resolution, capable of detecting smaller targets with greater precision. Signal processing algorithms, incorporating techniques like adaptive filtering and clutter suppression, can enhance the signal-to-noise ratio, allowing for the detection of weaker signals emanating from nano drones. However, even with these advancements, the limitations of radar remain; high-frequency radar often has a shorter range, and complex environments can still significantly degrade performance.

Infrared (IR) sensors offer another avenue for nano drone detection. These sensors detect the heat signatures emitted by the drones' motors and electronics. However, the small size of nano drones results in a correspondingly weak thermal signature, making detection challenging, particularly in environments with significant background thermal noise, such as urban areas or sunny conditions. To improve detectability, sophisticated IR cameras with high sensitivity and advanced image processing algorithms can be employed. These algorithms can enhance the contrast between the drone's thermal signature and the background, facilitating detection. The use of passive IR sensors, capable of detecting even slight temperature differences, offers another potential advantage, although their lower range is a trade-off. Combining IR detection with other sensor modalities is a highly effective strategy, as demonstrated in various counter-UAS systems.

Acoustic sensors present a further detection method. The motors and propellers of nano drones generate distinct acoustic signatures, albeit at low power levels. This necessitates the use of highly sensitive microphones and advanced acoustic signal processing techniques to isolate the drone's sounds from ambient noise. Acoustic localization techniques, using arrays of microphones to triangulate the source of the sound, can help pinpoint the drone's location. However, acoustic detection is highly susceptible to environmental noise, especially in urban environments, making it difficult to distinguish drone sounds from other sources of noise. Furthermore, the effectiveness of acoustic detection is highly dependent on the drone's operating mode; stealth drones, employing quiet propulsion methods, will be much more difficult to detect acoustically.

Beyond these individual sensor modalities, the use of data fusion is crucial for enhancing the effectiveness of nano drone detection. Data fusion involves combining information from multiple sensors to provide a more comprehensive and accurate picture of the situation. By integrating data from radar, IR, and acoustic sensors, along with potentially other sensor types like visual cameras or Lidar, a more robust and reliable detection system can be created. Advanced algorithms can correlate data from various sources, reducing false positives and improving the accuracy of drone localization. For instance, a thermal anomaly detected by an IR sensor can be corroborated by an acoustic signal detected by a microphone array in close proximity, resulting in increased confidence in the detection.

However, the challenge of detecting nano drones is exacerbated by their potential use of stealth technologies. These technologies aim to minimize the drone's detectability, making detection significantly more challenging. One form of stealth technology involves using materials with low radar cross-sections, reducing the signal reflected by the drone. Another technique focuses on minimizing the drone's thermal signature, potentially through advanced cooling systems or materials with low thermal emissivity. Finally, the implementation of quiet propulsion systems seeks to reduce the acoustic signature of the drone, making acoustic detection more challenging. Countering these stealth technologies requires employing more sophisticated detection systems with higher sensitivity and more advanced signal processing capabilities, pushing the boundaries of sensor technology and data fusion techniques.

The detection of nano drones in cluttered environments presents additional complexities. Urban canyons, dense forests, or complex industrial sites introduce significant amounts of clutter that can mask the signals generated by nano drones. This clutter can interfere with radar signals, mask thermal signatures, and drown out acoustic signals. Advanced signal processing techniques, including adaptive filtering, clutter suppression, and background subtraction, are essential for improving detection performance in cluttered environments. The use of multiple sensors with complementary capabilities, alongside sophisticated data fusion techniques, is crucial for maximizing detection probabilities and achieving accurate location estimates.

Furthermore, the potential for swarming behaviour significantly complicates detection and tracking. A swarm of nano drones can overwhelm individual sensor systems and confuse tracking algorithms. The small size and dispersed nature of the swarm make it challenging to track each individual drone. Countermeasures for swarm detection require the ability to detect the swarm as a collective entity, rather than focusing solely on individual drones. This may involve detecting the overall density or pattern of signals from multiple sensors and using advanced tracking algorithms capable of managing a large number of targets simultaneously. Developing robust algorithms that can differentiate a swarm from natural or artificial clutter presents a significant challenge in countering these emerging threats.

Finally, the effectiveness of any detection and tracking system hinges on the processing and analysis of the collected data. The sheer volume of data generated by multiple sensors, particularly in the context of a large nano drone swarm, requires powerful computing capabilities. Real-time data processing and analysis are crucial to provide timely alerts and support rapid response actions. Advanced algorithms, leveraging machine learning and artificial intelligence, are essential for automatically identifying and classifying nano drone signals, separating them from noise, and making accurate predictions about the drones' trajectories. The development of robust, reliable, and scalable data processing systems is critical to ensuring that these countermeasures are effective. The combination of sophisticated sensor technology, advanced signal processing, effective data fusion techniques, and real-time data analysis forms the cornerstone of an effective defense against the emerging threat of nano drones. The continuing development and refinement of these technologies will be paramount as these small but potentially disruptive technologies evolve.

Electronic warfare (EW) offers a powerful suite of countermeasures against nano drones, focusing on disrupting their communication and control systems. Jamming, a core component of EW, aims to overwhelm or corrupt the signals that guide these miniature aircraft, rendering them ineffective or causing them to malfunction. However, the effectiveness of jamming against nano drones presents a unique set of challenges and considerations.

Unlike larger drones that rely on more powerful and easily targeted communication signals, nano drones often utilize low-power, spread-spectrum communication protocols. These protocols are designed to be robust against interference and spread their signal across a wide range of frequencies, making them difficult to jam effectively. A simple, broad-spectrum jammer might overwhelm other legitimate communication systems, creating significant collateral damage. Therefore, sophisticated jamming techniques are needed, requiring a thorough understanding of the specific communication protocols employed by the targeted nano drones.

One approach involves employing frequency-hopping spread spectrum (FHSS) jamming. FHSS jammers rapidly switch between different frequencies, matching the hopping rate of the target drone's communication signal. This forces the drone to constantly adjust its reception, leading to signal degradation and potential loss of control. Adaptive jamming techniques are also being developed, where the jammer analyzes the drone's signal in real time, adjusts its jamming strategy to optimize interference, and learns to anticipate frequency hopping patterns. These adaptive systems can provide a more effective countermeasure against sophisticated nano drones employing complex communication protocols and anti-jamming techniques.

Another promising area is directed energy jamming. This approach uses focused beams of electromagnetic energy to target the drone's receiver directly, overwhelming its ability to receive control signals. This method offers greater precision than broad-spectrum jamming, minimizing the risk of interference with other communication systems. However, directed energy jammers require accurate targeting, sophisticated tracking systems, and potentially substantial power. The effectiveness of this method is directly tied to the accuracy of the targeting system and the ability to maintain a lock on the small target.

The development of anti-jamming techniques for nano drones also presents a considerable challenge. Drone manufacturers are constantly innovating, incorporating more robust communication protocols and employing strategies to mitigate the impact of jamming. These might include redundant communication channels, error-correcting codes, and advanced signal processing algorithms to filter out interference. This necessitates a continuous arms race in the development of jamming technologies, constantly pushing the boundaries of EW capabilities to stay ahead of emerging anti-jamming strategies.

Beyond single-drone jamming, a significant difficulty arises with nano drone swarms. The coordinated actions of a swarm amplify the difficulty of effective jamming. A single jammer may overwhelm some drones, but others may simply switch communication channels or rely on alternative control methods. To effectively counter swarms, a coordinated jamming network is required. This could involve multiple jammers working in concert, utilizing a distributed jamming strategy, adapting to the swarm's dynamic behaviour and quickly adjusting jamming frequencies and power levels to counter emerging patterns and mitigate the possibility of a swarm adapting to the jamming signals.

The limitations of jamming technologies must be acknowledged. The range of a jammer is limited by the power of the transmitter and the sensitivity of the receiver. The effectiveness of jamming is also affected by environmental factors such as terrain, atmospheric conditions, and the presence of other electromagnetic signals. Urban environments, densely populated with various communication devices, can significantly impede the effectiveness of jamming systems due to the dense electromagnetic spectrum already present. Furthermore, jamming can be easily countered by the simple implementation of a backup communication channel, making the drone resilient to jamming attacks.

There's also the critical consideration of unintended consequences. Broad-spectrum jamming can disrupt essential civilian communication systems, causing significant disruption and potentially endangering lives. Narrow-band jamming, while more precise, requires accurate identification of the drone's signal, and misidentification could lead to the disruption of critical infrastructure communications. Therefore, careful planning and targeted deployment are essential to mitigate these risks. The legal and ethical implications of jamming must also be thoroughly considered before deployment.

Beyond signal jamming, electronic warfare strategies against nano drones can also encompass other techniques. Spoofing, for example, involves sending false signals to the drone to deceive it, causing it to deviate from its intended course or land prematurely. This requires accurate understanding of the drone's navigation and control systems, allowing for the injection of misleading data into the drone's navigational inputs, causing it to become disoriented and lose control. Sophisticated spoofing techniques may also involve using other signals to mask or alter the drone's perception of its environment, rendering its navigation systems inaccurate and ineffective.

Another EW approach is data link exploitation. By intercepting and analyzing the communication signals between the drone and its operator or control system, valuable information can be gained about the drone's mission, location, and capabilities. This intelligence can be used to guide countermeasures or to pre-emptively disrupt the drone's operation. The ability to intercept and interpret these data links directly impacts the capability to effectively jam the signal or otherwise disrupts its operation.

Research is ongoing into developing more sophisticated jamming techniques specifically designed to overcome the challenges posed by nano drones. This includes exploring advanced signal processing algorithms, developing more powerful and energy-efficient jammers, and employing artificial intelligence to adapt jamming strategies in real-time. The development of AI-powered jamming systems promises to improve the adaptability and effectiveness of countermeasures against swarms, allowing for swift response and adaptability to evolving drone control methods and anti-jamming strategies.

The development of AI-driven countermeasures extends beyond simple jamming. AI could be used to predict the behaviour of a swarm, analyze its communication patterns, and optimize jamming strategies to maximize their effectiveness. Machine learning algorithms could analyze vast amounts of data from various sensors to identify patterns in drone behaviour, enabling pre-emptive countermeasures before the drones pose a threat. This predictive approach is critical in countering the rapidly evolving threat of nano drones.

Directed energy weapons (DEWs) represent a compelling countermeasure against the emerging threat of nano drones. Unlike electronic warfare techniques that disrupt communication signals, DEWs directly target the physical integrity of the drone, offering a potentially more decisive approach. Two primary types of DEWs are particularly relevant in this context: lasers and high-power microwaves (HPM).

Lasers, with their highly focused beams of coherent light, can offer precise targeting capabilities. The energy delivered by the laser beam can generate sufficient heat to damage or destroy critical components of the nano drone, such as its onboard electronics, motors, or batteries. However, the effectiveness of laser countermeasures is significantly influenced by several factors. Firstly, the power output of the laser is central. Low-power lasers may only be capable of momentarily blinding the drone's sensors or causing minor damage, while high-power lasers are necessary for complete incapacitation or destruction. The challenge lies in balancing the power required for effectiveness with the constraints of size, weight, and power consumption, particularly for portable or mobile DEW systems.

Another critical factor is the accuracy of the targeting system. Nano drones are incredibly small, and precise targeting is essential to ensure the laser energy is effectively delivered to the intended target. This requires advanced tracking and pointing systems capable of compensating for atmospheric effects like turbulence, wind, and target movement. The ability to maintain a lock on a rapidly manoeuvring nano drone is essential for successful engagement. Advanced sensor fusion techniques, combining data from multiple sensors such as infrared cameras and radar, can enhance targeting accuracy. Furthermore, the development of autonomous targeting systems employing AI algorithms is crucial for efficiently neutralizing swarms of nano drones. These AI systems can quickly identify and prioritize targets, track their movements, and adjust laser aiming parameters in real-time.

The wavelength of the laser also plays a crucial role in its effectiveness. Different wavelengths have varying degrees of absorption by different materials. Selecting an optimal wavelength is essential to maximize the damage inflicted on the drone's components while minimizing collateral damage to the surrounding environment. For instance, certain wavelengths might be particularly effective at damaging specific materials commonly used in drone construction, while others might be more readily absorbed by atmospheric constituents. Research into the optimal laser wavelengths for targeting various nano drone materials and designs is an ongoing area of development.

Beyond the technical challenges, the use of laser DEWs raises significant safety and ethical considerations. The potential for accidental harm to humans or other sensitive equipment in the vicinity of the laser beam necessitates stringent safety protocols and precise control mechanisms. Safety interlocks and fail-safe systems are imperative to prevent accidental activation or misdirected beams. The potential for eye damage due to laser exposure necessitates careful consideration of both operational procedures and the design of the DEW system itself.

High-power microwaves (HPM) offer a different approach to neutralizing nano drones. HPM systems generate intense bursts of electromagnetic energy that can disrupt the drone's electronic components, causing malfunctions or complete system failure. Unlike lasers, HPM weapons don't rely on precisely focused beams, making them potentially more effective against swarms of rapidly moving drones. However, their effectiveness is also dependent on the power output and the susceptibility of the drone's electronics to electromagnetic interference. The development of nano drones with increased electromagnetic shielding or robust electronics is an ongoing challenge that may reduce the effectiveness of HPM weapons.

The power requirements of HPM weapons are substantial, posing logistical challenges for deployment in certain scenarios. Developing compact and portable HPM systems with sufficient power output to effectively neutralize nano drones is an active area of research and development. Advances in power generation technologies, particularly in high-power, lightweight power sources, are crucial for making HPM weapons more practical and deployable. Moreover, the effectiveness of HPM weapons can be affected by environmental factors, such as the presence of other electromagnetic interference or atmospheric conditions.

Similar to lasers, HPM weapons also pose safety concerns. Exposure to intense microwave radiation can have harmful effects on human health and sensitive electronic equipment. Strict safety protocols and controlled deployment are crucial to minimize the risks associated with HPM technology. The development of advanced safety features, such as directional beams to limit radiation spread and automated safety shutoff mechanisms, is vital for responsible and safe operation.

Further research is needed to optimize the design and deployment of both laser and HPM DEWs for maximum effectiveness against nano drones. This includes developing more efficient energy sources, improving targeting accuracy, and mitigating the potential safety and environmental risks. Furthermore, the development of counter-countermeasures by nano drone manufacturers necessitates ongoing research and development of DEW systems to maintain their effectiveness. This is an ongoing arms race between the developers of nano drone technology and those developing countermeasures.

The development of hybrid DEW systems, combining the precise targeting capabilities of lasers with the area-of-effect capabilities of HPM systems, could potentially offer a more effective approach to countering nano drone swarms. Such systems could allow for the prioritization and neutralization of high-value targets with lasers while simultaneously disabling or disrupting a larger number of drones with HPM beams. This strategic combination of technologies may offer the most comprehensive approach to neutralizing this evolving threat. Furthermore, the integration of AI-driven targeting and control systems will greatly enhance the effectiveness and autonomy of DEW systems in combating swarms of nano drones.

Beyond the technological aspects, legal and ethical considerations must be carefully addressed. The potential for unintentional harm to civilians and the risks of escalating conflicts necessitate a comprehensive framework for the responsible use and deployment of DEW systems. International agreements and regulations are essential to ensure that the use of these powerful weapons is governed by strict safety and ethical standards. Transparency and accountability in the development and deployment of DEW technology are crucial to maintaining international security and preventing potential misuse.

Physical countermeasures offer a tangible, albeit often less sophisticated, approach to neutralizing nano drones. Unlike directed energy weapons, which rely on the precise application of energy to disable or destroy the drone, physical methods focus on intercepting or trapping the drone through physical means. This approach, while seemingly simple, presents a unique set of challenges and opportunities depending on the specific design and deployment scenario.

One of the most widely considered physical countermeasures is the use of nets. These nets, often deployed from handheld launchers or mounted on vehicles, aim to physically entangle the nano drone, preventing it from further flight and potentially capturing it for analysis. The design of these nets is crucial. The mesh size must be fine enough to capture the miniature drone, yet robust enough to withstand the impact of the drone and the stresses of deployment. Materials must also be selected with consideration for the environment. In outdoor settings, the net material must be resistant to weathering and degradation from sunlight, rain, and temperature extremes. The net must also be sufficiently lightweight and easily deployable to be effective in fast-paced scenarios.

Beyond the material properties, the deployment mechanism is equally important. A rapid and accurate deployment is vital to successfully capture a nano drone, particularly when dealing with multiple drones or drones exhibiting high manoeuvrability. Electromagnetic or pneumatic launchers offer high-speed deployment, although the complexity and potential for failure need careful consideration. Simpler mechanical launching systems offer increased reliability, but may be slower and less suitable for rapid response scenarios. Furthermore, the size and weight of the launcher itself needs careful consideration, as a large and unwieldy system limits deploy ability and tactical flexibility.

Traps represent another category of physical countermeasures. These traps can range from simple adhesive surfaces to more complex electromechanical devices designed to ensnare or disable nano drones. Adhesive traps, for example, could be deployed as patches on sensitive infrastructure or integrated into buildings. The adhesive must be sufficiently strong to hold the nano drone, but not so strong as to cause damage or be difficult to remove. Furthermore, the adhesive should be selectively designed to avoid attracting unintended objects or creating environmental hazards.

More sophisticated electromechanical traps could incorporate sensors to detect the presence of a nano drone and then activate a mechanism to capture or disable it. These could involve intricate arrays of micro-mechanisms or cleverly designed barriers, potentially using electrostatic forces or carefully positioned obstacles. The design of these traps is paramount, requiring a balance between effectiveness, reliability, and the ability to prevent accidental triggering. Environmental factors, such as wind or precipitation, could also impact the performance of these traps, highlighting the need for robust and resilient designs.

Beyond nets and traps, other physical barriers can also play a role in mitigating the threat of nano drones. These might include specialized meshes or screens integrated into building facades or other sensitive infrastructure. These barriers need to be designed considering the size of the nano drone and its potential flight capabilities. A robust material selection, capable of withstanding high impact forces and environmental stresses, is essential. The deployment of such barriers could involve substantial cost and logistical challenges, particularly for large-scale applications.

However, the effectiveness of physical countermeasures faces significant limitations. The sheer number of drones in a swarm significantly reduces the effectiveness of individual nets or traps. A swarm attack would likely overwhelm any single physical countermeasure, requiring a vastly scaled-up system of interconnected nets, traps, and barriers, a solution with substantial logistical and cost implications. Furthermore, advanced nano drone designs might incorporate evasion techniques, utilizing highly agile flight patterns or sensor jamming to avoid physical interception. Advanced materials, like carbon nano tubes or other high-strength, low-weight materials, might increase the drone's resistance to physical capture.

The development of counter-countermeasures also represents a significant challenge. Manufacturers of nano drones could develop designs that specifically circumvent these physical barriers, such as drones with enhanced manoeuvrability, improved structural integrity to withstand impact, or even drones equipped with countermeasures that physically disable the nets or traps. This represents an ongoing arms race between those developing the drones and those developing countermeasures.

Another significant limitation is the difficulty in scaling physical countermeasures to handle large-scale swarm attacks. Deploying a sufficient number of nets or traps to effectively counter a massive swarm would be a logistical nightmare, requiring an immense investment in resources and personnel. The cost of deployment and maintenance of these systems would potentially outweigh the benefits, particularly in situations where the value of the protected assets is relatively low.

Cyber security presents a decisive, yet often overlooked, layer of defence against nano drone threats. While physical countermeasures offer a tangible approach to neutralization, they often fall short against swarms or drones equipped with sophisticated evasion techniques. Targeting the software and control systems of these miniature machines opens up a potent avenue for countermeasures, offering the potential for disruption, control, or complete neutralization without the limitations of physical interception. This software-based approach, however, introduces a complex landscape of ethical and security implications that must be carefully navigated.

One of the most direct methods involves exploiting vulnerabilities in the nano drone's software. Just as larger systems are susceptible to hacking, nano drones, despite their size, are vulnerable to software exploits. These exploits can range from simple buffer overflows – leveraging memory management flaws to inject malicious code – to more sophisticated attacks that target specific functionalities within the drone's operating system or application layers. Sophisticated reverse engineering techniques, often requiring the physical acquisition of a drone for detailed analysis, are necessary to identify these vulnerabilities. Once vulnerability is identified, malicious code can be crafted and deployed, potentially leading to a variety of outcomes, ranging from simple disruptions to complete control.

For instance, a successful exploit could disable the drone's navigation system, causing it to lose its way and become effectively useless. Alternatively, it could manipulate the drone's flight parameters, causing it to crash or land prematurely. More insidious attacks could involve hijacking the drone's control system, allowing an attacker to remotely pilot the device and redirect it to a different target or gather sensitive information. The potential for such attacks to compromise sensitive data collected by the drone – such as high-resolution imagery or audio recordings – is a significant concern.

The use of malware presents another potent cyber security countermeasure. Malware specifically designed to target nano drone software could be deployed through a variety of methods, including network infections (if the drones communicate wirelessly), through software updates, or even physically via a compromised component. This malware could perform various malicious functions, such as corrupting the drone's firmware, disabling key components, or even establishing a backdoor for remote control. The design of such malware is a complex undertaking, requiring detailed knowledge of the drone's operating system, communication protocols, and hardware architecture. The challenge lies in developing malware that is effective against a wide range of drone models and software versions, while minimizing the risk of detection.

Furthermore, the development of anti-malware systems for nano drones poses a considerable challenge. The limited processing power and memory capacity of these miniature devices restricts the range of defensive software that can be implemented. Traditional antivirus techniques may not be applicable, requiring innovative approaches tailored to the unique constraints of nano drone systems. This is a continuously evolving arms race, with drone manufacturers developing more robust security measures and researchers striving to develop increasingly sophisticated attack vectors.

Software-defined radio (SDR) jamming presents a different type of cyber security countermeasure. This technology utilizes software-controlled radio transmitters to generate jamming signals targeted at the specific communication frequencies used by nano drones. By overwhelming the drone's communication channels with noise or deceptive signals, the SDR can disrupt the drone's ability to communicate with its operator or receive instructions. This approach offers a wide-area denial-of-service effect, effectively grounding a large number of drones without necessarily targeting individual devices. The effectiveness of SDR jamming depends on several factors, including the power of the jammer, the range of its broadcast, and the resilience of the drone's communication system. Sophisticated drones may employ frequency hopping or other anti-jamming techniques, necessitating a constant evolution of jamming strategies to maintain effectiveness. Moreover, the development of sophisticated, adaptive jamming systems is needed to counter advanced signal processing in the drones. The use of AI in identifying and responding to drone communication patterns can significantly enhance the effectiveness of these systems.

The ethical and security implications of employing these cyber security countermeasures are profound. The use of hacking, malware, or jamming techniques against nano drones, especially in military contexts, raises concerns about unintended consequences and potential escalation. Accidental damage to civilian infrastructure or unintended interference with other wireless devices are significant possibilities. Moreover, the use of these methods against autonomous weapons systems raises questions about international law and the responsibility for any harm inflicted. The development and deployment of such countermeasures necessitate the establishment of clear ethical guidelines and robust security protocols to minimize the risk of unintended consequences. This includes thorough testing and validation of these systems to ensure that they only target the intended nano drones and do not cause harm to innocent parties or essential infrastructure. International agreements and regulations are vital to establish acceptable standards of usage, preventing an escalation of cyber warfare.

Furthermore, the possibility of these cyber security countermeasures being weaponized against civilian drones or other critical infrastructure requires careful attention. The potential misuse of these technologies for malicious purposes necessitates the development of counter-countermeasures to protect against attacks and the implementation of robust safeguards to prevent unauthorized access and control. This underlines the need for robust cyber security legislation and regulations that clearly define acceptable usage and penalize malicious activity. The establishment of international cooperation and information sharing frameworks is crucial to mitigate the risks associated with the proliferation of such technologies. Regular auditing of the software and hardware used in nano drones, along with the development of secure communication protocols and robust authentication mechanisms, is essential to maintain the integrity and safety of drone operations.

The development of secure and reliable operating systems specifically designed for nano drones is crucial. These operating systems should incorporate advanced security features, such as secure boot processes, cryptographic authentication, and intrusion detection systems. Regular updates and patches to address newly discovered vulnerabilities are essential for maintaining a high level of security. The use of hardware-based security measures, such as secure elements or trusted execution environments, can provide an additional layer of protection.

Nano Weapons

The convergence of nanotechnology and weaponry represents a paradigm shift in the design and capabilities of advanced arms. Nanotechnology, the manipulation of matter at the atomic and molecular level, offers unprecedented opportunities to enhance existing weapon systems and create entirely new classes of weaponry. The unique properties of nano materials, stemming from their extremely small size and high surface area-to-volume ratio, are the driving force behind this revolution. These properties include exceptional strength-to-weight ratios, enabling the creation of lighter, more durable weapons; enhanced reactivity, leading to more powerful explosives and propellants; and tunable properties, allowing for the precise tailoring of material characteristics to specific weapon applications.

For instance, the high tensile strength of carbon nano tubes, a cylindrical form of carbon atoms, makes them ideal for reinforcing structural components in weapons, resulting in increased resilience to damage and improved performance under stress. Graphene, a single layer of carbon atoms arranged in a hexagonal lattice, exhibits exceptional electrical conductivity, making it suitable for enhancing the efficiency of directed energy weapons (DEWs) such as lasers. Similarly, metal nano particles, owing to their high surface area, can significantly increase the reactivity of explosives and propellants, leading to more powerful and efficient weapons. The ability to control the size, shape, and surface chemistry of these nano particles further allows for fine-tuning their properties, optimizing their performance for specific applications.

The integration of nanotechnology into existing defence applications is already underway. Nano materials are currently used in various military technologies, such as advanced armour, lightweight composite materials for aircraft and vehicles, and improved sensors for surveillance and targeting. For example, nano structured ceramics are being incorporated into body armor to enhance its ballistic protection capabilities. Nano materials are also being explored for use in creating stealth technologies, reducing the radar and infrared signatures of military vehicles and aircraft. These advancements showcase the potential of nanotechnology to enhance existing military capabilities and pave the way for more transformative applications in the future.

However, the development and deployment of nanotechnology-enhanced weapons raise complex ethical considerations. The potential for misuse and unintended consequences require careful consideration. The increased lethality and precision of these weapons, coupled with their potential for miniaturization, pose significant challenges for international security and arms control. The development of autonomous weapons systems incorporating nanotechnology further complicates the ethical landscape, raising concerns about accountability and the potential for unintended escalation of conflicts.

The potential for proliferation also presents a major concern. The relatively low cost and ease of production of some nano materials could make the technology accessible to non-state actors and rogue nations, increasing the risk of global instability. Ensuring that the development and deployment of nanotechnology-enhanced weapons are guided by ethical principles and international norms is therefore crucial. This necessitates a robust international framework for the control and regulation of nanotechnology in the military domain. Collaboration between governments, research institutions, and civil society is essential to develop effective safeguards and mitigate the potential risks.

The unique properties of nano materials have far-reaching implications for the design and functionality of nuclear and related high-energy systems. In boosted fission weapons, nano materials can significantly improve fuel containment and energy release. Boosted fission weapons enhance the efficiency of nuclear fission by incorporating a small amount of fusion fuel, typically deuterium and tritium, into the weapon core. The fusion reaction amplifies the fission process, resulting in a more powerful explosion with less fissile material. Nano materials play a crucial role in ensuring the efficient mixing and ignition of the fusion fuel.

The extremely high temperatures and pressures generated during the fusion reaction require robust containment mechanisms. Nano materials, with their high strength-to-weight ratio and tunable properties, can be used to design more durable and efficient containment systems. For example, nano structured ceramics and metallic glasses can withstand the extreme conditions better than conventional materials, enhancing the weapon's efficiency and safety. Furthermore, nano scale control of the fuel mixture can optimize the energy release, potentially leading to more compact and powerful weapons.

The potential of nanotechnology extends beyond boosted fission weapons to the realm of pure fusion weapons, a far more ambitious endeavor. Pure fusion weapons, which would rely solely on fusion reactions without the need for a fission trigger, represent the ultimate goal of nuclear weapon miniaturization. However, achieving controlled and sustained fusion reactions without fission triggers presents immense technological challenges. The key obstacles include plasma confinement, the need to contain the extremely hot and energetic plasma generated during the fusion process, and achieving ignition, the point at which the fusion reaction becomes self-sustaining.

Nanotechnology offers several potential avenues for overcoming these challenges. For example, nano structured materials could be used to create more efficient and durable confinement systems capable of withstanding the extreme temperatures and pressures of a fusion reaction. Nano materials could also enhance the efficiency of inertial confinement fusion (ICF) techniques, a promising approach to achieving controlled fusion. In ICF, a small pellet of fusion fuel is compressed to extremely high densities and temperatures using powerful lasers or particle beams. Nano materials could enhance the laser-energy coupling efficiency, enabling more efficient compression of the fusion fuel.

Directed energy weapons (DEWs), such as high-powered lasers and electromagnetic pulse (EMP) devices, are another area where nanotechnology is poised to revolutionize military capabilities. Nano materials can significantly enhance the energy control, miniaturization, and efficiency of these weapons. For example, nano structured materials can improve the heat dissipation capabilities of laser systems, enabling them to operate at higher power levels. Nano materials can also enhance the energy storage density of batteries and capacitors used to power DEWs, reducing the size and weight of these weapons.

The development of more compact and efficient EMP devices could revolutionize electronic warfare. Nano materials can potentially enhance the effectiveness of EMP generators by improving the efficiency of energy conversion and miniaturizing the device's components. This could lead to more portable and readily deployable EMP weapons capable of disrupting enemy communication and electronic systems.

In addition to enhancing existing weapon systems, nanotechnology opens up the possibility of developing entirely new classes of weaponry. Antimatter weapons, for example, theoretically possess the highest energy density of any known substance. However, the safe storage and control of antimatter represent formidable technological challenges. Nanomaterials could play a crucial role in creating stable containment systems capable of holding antimatter for extended periods.

Nano thermites, highly reactive mixtures of nano scale metal particles and oxidizers, offer another avenue for developing advanced weaponry. The extremely high surface area of the nano particles results in much faster and more energetic reactions compared to conventional thermites. Nano thermites could be used to create more powerful and efficient explosives and propellants. Plasma weapons, which utilize highly ionized gases, or plasma, as the primary destructive agent, are also a promising area for the application of nanotechnology. Nano materials could enhance the generation, control, and focusing of plasma streams, leading to more effective and versatile weapons.

The development and deployment of these advanced weapon systems will, however, require meticulous attention to safety and security protocols. Rigorous testing and evaluation are essential to ensure the reliability, safety, and effectiveness of nanotechnology-enhanced weapons. Careful consideration must be given to the potential environmental and health risks associated with these technologies. International cooperation and arms control agreements are vital to prevent the proliferation of these powerful weapons and ensure their responsible use. The convergence of nanotechnology and weaponry is a double-edged sword, offering immense potential for both advancement and destruction, underscoring the necessity for responsible innovation and careful ethical consideration.

The remarkable potential of nanotechnology in revolutionizing weaponry stems directly from the exceptional properties of nano scale materials. These materials, defined by at least one dimension measuring less than 100 nano meters (a nano meter being one billionth of a meter), exhibit characteristics drastically different from their bulk counterparts. This difference arises from the significant increase in surface area relative to volume at the nano scale, leading to enhanced reactivity, unique mechanical properties, and tailored electronic behaviour.

One prominent example is carbon nano tubes (CNTs), cylindrical structures of carbon atoms arranged in a hexagonal lattice. Their exceptional tensile strength, exceeding that of steel by several orders of magnitude, makes them ideal candidates for reinforcing structural components in weapons systems. These results in lighter, yet far stronger, weaponry with improved resistance to damage and enhanced performance under extreme conditions, such as the stresses experienced during ballistic impact or high-velocity launch. Furthermore, their high aspect ratio (length-to-diameter ratio) allows for the creation of exceptionally strong and lightweight fibbers, suitable for a wide range of applications, from reinforcing the casings of projectiles to creating ultra-strong yet flexible composite materials for aircraft and armoured vehicles.

Graphene, a single layer of carbon atoms arranged in a hexagonal lattice, possesses equally remarkable properties. Its exceptional electrical conductivity, surpassing that of copper, makes it a prime candidate for improving the efficiency of directed energy weapons (DEWs), particularly high-powered lasers. The ability of graphene to efficiently conduct and dissipate heat is crucial in high-energy laser systems, allowing for higher power output and longer operational times without overheating. Moreover, graphene's flexibility and transparency could lead to the development of innovative, lightweight optical components for laser systems, potentially miniaturizing these weapons and enhancing their mobility.

Metal nano particles, with their extremely high surface area-to-volume ratios, exhibit enhanced reactivity compared to their bulk counterparts. This increased reactivity translates directly to more powerful and efficient explosives and propellants. For instance, aluminium nano particles are currently being investigated for their potential to improve the performance of solid rocket propellants, enhancing their burn rate and energy output. The ability to precisely control the size and shape of these nano particles allows for fine-tuning of their reactivity, leading to tailored explosive characteristics such as increased detonation velocity or adjusted energy release profiles.

Beyond these widely studied materials, a vast array of other nano scale materials offer unique possibilities for advanced weaponry. Nano structured ceramics, for example, exhibit exceptional hardness, high-temperature resistance, and wear resistance, making them ideal for applications requiring extreme durability. These materials are already finding their way into advanced body armor, providing enhanced ballistic protection. Similarly, metallic glasses, amorphous metallic alloys, possess unique combinations of strength, ductility, and corrosion resistance, making them suitable for components requiring high strength and resilience in harsh environments.

The synthesis and characterization of these nano materials are critical to realizing their full potential in weaponry. Various techniques are employed for their production, including chemical vapor deposition (CVD) for CNTs and graphene, sol-gel methods for metal oxides, and sputtering techniques for metal nano particles. The precise control over the synthesis process is crucial in determining the final properties of the nano materials, including their size, shape, surface chemistry, and defect density. Advanced characterization techniques, such as transmission electron microscopy (TEM), scanning electron microscopy (SEM), X-ray diffraction (XRD), and various spectroscopic methods, are employed to analyze the structure, morphology, and composition of the synthesized nano materials.

However, the production and application of nano materials in weaponry are not without their challenges. The synthesis of high-quality nano materials often requires specialized equipment and expertise, resulting in potentially high production costs. Furthermore, controlling the uniformity and reproducibility of nano material properties across large scales remains a significant challenge. Ensuring consistent performance and reliability of weapon systems incorporating nano materials requires a thorough understanding of the material's behaviour under various conditions, including extreme temperatures, pressures, and mechanical stresses.

Another crucial aspect is the potential toxicity of certain nano materials. The high surface area of nano particles can lead to increased reactivity and potential health risks. Rigorous testing and evaluation are essential to assess the potential environmental and health implications of nano materials employed in weaponry, ensuring the safety of both personnel handling these materials and the environment. This requires strict adherence to safety protocols and the development of mitigation strategies to minimize potential risks. The long-term effects of exposure to various nano materials also require extensive study to ensure responsible and safe implementation of this transformative technology.

Moreover, the ethical considerations surrounding the development and deployment of nanotechnology-enhanced weapons cannot be overlooked. The potential for misuse and proliferation of these technologies necessitates careful consideration of the societal implications. The increased lethality and precision of nanotechnology-enhanced weapons raise significant concerns about the potential for escalation of conflicts and the erosion of existing arms control agreements. International cooperation and stringent regulatory frameworks are essential to ensure the responsible development and deployment of these advanced technologies, mitigating the risks associated with their potential for misuse.

The development of self-replicating nano bots, while still largely theoretical, presents a particularly alarming scenario. The potential for uncontrolled self-replication, leading to a "grey goo" scenario, where nano bots consume all matter in their path, represents a significant concern. This highlights the urgent need for rigorous safety protocols and ethical guidelines to govern the development and application of nanotechnology, ensuring its benefits are harnessed while preventing catastrophic consequences. This necessitates a multi-faceted approach involving scientists, policymakers, and the public to ensure the responsible stewardship of this potentially revolutionary technology. Ongoing research and careful consideration of potential risks are essential to navigate the complex ethical and societal challenges associated with this transformative field. The future of nanotechnology in weapons hinges on our ability to harness its potential while proactively mitigating its inherent risks.

The integration of nanotechnology into nuclear weapon design represents a significant leap forward, particularly in the realm of boosted fission weapons. These weapons, while still relying on the fission of heavy elements like uranium or plutonium for their primary explosive yield, leverage a secondary fusion reaction to dramatically enhance their efficiency. This enhancement stems from the fusion process's ability to generate copious neutrons, which in turn, significantly increase the fission chain reaction within the primary fission stage. The result is a smaller, lighter, and considerably more powerful weapon compared to a purely fission-based device. However, the critical challenge lies in achieving efficient mixing and ignition of the fusion fuel – typically deuterium and tritium – within the extremely short timescale of the implosion process. This is where nanotechnology plays a transformative role.

Traditionally, boosted fission weapons relied on gaseous deuterium-tritium (DT) mixtures carefully contained within the weapon's core. However, the precise control and efficient mixing of the gas during the implosion process presented significant challenges. Nanotechnology offers a pathway to overcome these limitations through the use of nano structured materials to improve fuel containment, ignition, and the overall efficiency of the boosted fission process. One promising approach involves the use of nano engineered foams, consisting of a network of interconnected nano-ized pores filled with the DT fuel. These foams offer a high surface area-to-volume ratio, facilitating rapid and complete mixing of the fuel during implosion. The increased surface area also enhances the interaction between the fusion fuel and the high-energy neutrons produced during fission, significantly increasing the rate of fusion reactions. The precise control over the pore size and distribution within these nano foams allows for fine-tuning of the fuel's burn characteristics and maximizing the overall energy yield.

Another critical aspect is the introduction of nano scale catalysts that facilitate the fusion reactions. The fusion process, while energetically favourable, requires surmounting a significant energy barrier for the reaction to initiate. Nano particles of certain metals, such as palladium or platinum, are known to possess catalytic properties that can lower this activation energy, effectively increasing the fusion reaction rate. The use of these nano particles, strategically incorporated into the nano foam structure, could significantly enhance the efficiency of the boosted fission process. The precise distribution and concentration of these catalytic nano particles are crucial to optimize their effectiveness and ensure a controlled and efficient fusion burn.

Furthermore, nano materials can play a vital role in improving the structural integrity and durability of the weapon's components. The extreme conditions encountered during the implosion – pressures exceeding several million atmospheres and temperatures reaching tens of millions of degrees Celsius – demand materials with exceptional strength, hardness, and resistance to extreme environments. Nano structured ceramics, such as those based on boron carbide or silicon carbide, possess superior mechanical properties compared to their conventional counterparts. These materials, incorporating nano sized reinforcing particles, can withstand the tremendous stresses involved in the implosion process, ensuring the integrity of the weapon's casing and preventing premature detonation or malfunction. Moreover, the improved thermal conductivity of these nano materials could enhance the dissipation of heat generated during the fusion process, contributing to a more controlled and efficient reaction.

The use of nano tubes, either carbon nano tubes (CNTs) or boron nitride nano tubes (BNNTs), is also under investigation for enhancing the performance and reliability of boosted fission weapons. These materials' exceptional strength and flexibility can be leveraged to reinforce the structural components of the weapon, improving its resistance to damage and extending its operational lifespan. Their high aspect ratio and exceptional tensile strength provide a significant advantage over traditional reinforcement materials. The incorporation of nano tubes into the weapon's core could also improve the uniformity of the implosion process, leading to a more complete and efficient fusion reaction.

However, the introduction of nano materials into nuclear weapon design presents significant challenges in ensuring safety and reliability. The potential toxicity of certain nano particles requires meticulous attention, demanding rigorous testing and evaluation to assess their potential health and environmental impacts. Furthermore, the long-term stability and durability of nano materials under extreme conditions need to be thoroughly investigated to ensure the weapon's reliable performance over its entire operational lifespan. The potential for material degradation or unexpected reactions under extreme pressures and temperatures could compromise the weapon's functionality and potentially lead to catastrophic failures.

The precise control and uniformity of the nano materials' properties are also crucial. Even minor variations in the size, shape, or surface chemistry of nano particles can drastically alter their reactivity and effectiveness. Ensuring consistent and reproducible synthesis and assembly of the nano materials is essential for guaranteeing the weapon's reliability and predictability. This necessitates the development of advanced manufacturing techniques capable of producing high-quality nano materials at scale while maintaining strict control over their properties. Advanced characterization techniques, such as electron microscopy, X-ray diffraction, and various spectroscopic methods, are vital in ensuring quality control and verifying the desired properties of the nano materials before integration into the weapon system.

Beyond the materials science aspects, the design and integration of these nano engineered components into the weapon system require advanced simulation and modelling capabilities. Computational fluid dynamics (CFD) simulations can be used to predict the behaviour of the nano foam fuel during implosion, optimizing its design to maximize the fusion reaction rate and the overall energy yield. Detailed numerical simulations are crucial in evaluating the weapon's performance under different conditions and identifying potential design flaws. Moreover, these simulations are crucial for understanding the impact of variations in nanomaterial properties on the overall weapon performance.

The potential for improved energy release and enhanced efficiency in boosted fission weapons through the application of nanotechnology is undeniable. However, a multifaceted approach, integrating advanced materials science, sophisticated computational modeling, and rigorous safety testing, is crucial for realizing the full potential of this technology while mitigating the potential risks. The responsible development and deployment of these advanced weapons systems demand careful consideration of safety, reliability, and ethical implications, ensuring that the technological advancements are harnessed for peaceful purposes and deployed responsibly, avoiding any potential misuse or escalation of conflicts. Continuous research and development in this field are paramount to achieving the balance between enhancing nuclear weapon technology's capabilities and maintaining global security and stability.

The discussion thus far has centred on the application of nanotechnology to enhance boosted fission weapons, leveraging the fusion process to amplify the explosive yield of a fission primary. However, the ultimate goal in weapons design, from a purely energetic perspective, is to achieve a purely fusion-based weapon—a device that generates explosive power solely from fusion reactions, eliminating the need for fission trigger entirely. This presents a significantly greater challenge. Pure fusion weapons, also sometimes referred to as "clean" thermonuclear weapons (although the term "clean" is misleading regarding the radiological fallout), remain largely theoretical, at least in terms of practically deployable devices. The inherent difficulties in achieving controlled, self-sustaining fusion reactions without the assistance of a fission primary are formidable.

The core problem lies in the extreme conditions required to initiate and sustain a fusion reaction. Fusion, the process of combining light atomic nuclei to form heavier ones, releases vast amounts of energy but only under conditions of incredibly high temperature and pressure. These conditions are necessary to overcome the Coulomb barrier—the electrostatic repulsion between positively charged nuclei. In boosted fission weapons, the fission primary provides both the necessary temperature and neutron flux to initiate the fusion reaction. In a pure fusion weapon, however, this crucial initial energy input must be generated without the assistance of a fission process. This necessitates incredibly powerful and precise energy delivery mechanisms.

One of the most promising approaches to pure fusion weapon design is inertial confinement fusion (ICF). In ICF, a small pellet of fusion fuel, typically deuterium and tritium, is imploded to extremely high densities and temperatures using high-powered lasers or particle beams. The implosion compresses the fuel to conditions where fusion reactions can occur, resulting in a significant energy release. However, achieving the necessary level of implosion symmetry and energy delivery precision to initiate a self-sustaining fusion reaction remains an enormous challenge. Even small asymmetries in the implosion can lead to premature fuel disassembly and a failure to achieve the critical conditions for fusion.

Nanotechnology offers potential avenues for improving the efficiency and reliability of ICF. For instance, precisely engineered nano structured fuel pellets could enhance the implosion process. The fuel pellet's surface properties, particularly its uniformity and smoothness, significantly affect the symmetry of the implosion. Nano engineered surfaces with precisely controlled roughness and composition can minimize instabilities during the implosion, improving energy coupling and enhancing the probability of achieving the required conditions for sustained fusion. Furthermore, embedding nano sized catalysts within the fuel pellet could lower the activation energy needed to initiate the fusion process, increasing the efficiency of the reaction. These catalysts could be strategically positioned within the fuel to maximize their effectiveness.

Another critical aspect of ICF is the energy delivery mechanism. High-powered lasers or particle beams are required to implode the fuel pellet with sufficient precision and energy. Nanotechnology could potentially play a role in improving the efficiency and miniaturization of these energy sources. For instance, nano structured optical components could enhance laser energy efficiency and beam quality. Similarly, nano materials with high thermal conductivity could improve the heat dissipation from high-power systems, enhancing their reliability and longevity.

Beyond ICF, other approaches to pure fusion are under investigation. Magnetic confinement fusion (MCF), typically employed in experimental fusion reactors like tokamaks, relies on powerful magnetic fields to confine a plasma of deuterium and tritium, heating it to temperatures where fusion can occur. While MCF shows promise for energy generation, its application in a weapon context faces significant challenges. The scale and complexity of MCF devices make them unsuitable for weapons in the foreseeable future. However, nano materials could potentially enhance the performance of MCF systems, for example, by improving the magnetic confinement properties or the heat dissipation from the plasma. Nano structured materials with enhanced magnetic properties might improve the effectiveness of the magnetic confinement fields, while materials with high thermal conductivity could help manage the immense heat generated in the plasma.

Despite the theoretical potential, the development of a practical pure fusion weapon faces immense technological hurdles. The required energy densities for self-sustaining fusion are orders of magnitude greater than what is currently achievable with even the most advanced lasers or particle accelerators. The extreme temperatures and pressures involved demand materials with unprecedented properties, pushing the limits of materials science. The precision and control required for the implosion process or plasma confinement are beyond the capabilities of current technology. Even minor imperfections in the fuel pellet, implosion symmetry, or plasma confinement could lead to failure to achieve the necessary conditions for a significant fusion yield.

Beyond the purely scientific and engineering challenges, there are significant hurdles related to safety and reliability. The handling, transport, and deployment of a pure fusion weapon would present unique logistical and safety concerns. The extreme energy densities involved necessitate robust containment measures to prevent accidental detonation. Furthermore, the potential for the release of radioactive by-products from a malfunctioning device poses serious environmental and health hazards.

The development of a pure fusion weapon remains a distant prospect. While nanotechnology offers potential solutions to some of the key challenges, significant breakthroughs are still required to overcome the fundamental limitations of achieving controlled, self-sustaining fusion reactions without a fission primary. The research and development required are extensive and costly, with no guarantee of success. However, the potential payoff—a potentially more efficient and "cleaner" nuclear weapon—is enough to drive continued research in this field. Nevertheless, the ethical implications of developing even more powerful nuclear weapons should be carefully considered, with international efforts towards disarmament and non-proliferation remaining crucial. The potential benefits must be carefully weighed against the inherent dangers and the long-term consequences for global security. The pursuit of this technology must be guided by responsible scientific ethics and a commitment to peace and security.

The transition from discussing the challenges of pure fusion weapons to the realm of directed energy weapons (DEWs) might seem abrupt, but the underlying theme remains consistent: the relentless pursuit of enhanced energy control and miniaturization in advanced weaponry. Nanotechnology, with its ability to manipulate matter at the atomic and molecular level, offers unprecedented opportunities to revolutionize DEWs, making them more powerful, efficient, and compact. This section will explore the application of nanotechnology to two key types of DEWs: high-powered lasers and electromagnetic pulse (EMP) devices.

High-powered lasers, capable of delivering intense beams of light energy over long distances, have been a subject of military research for decades. Their potential applications range from precision targeting and destruction of enemy assets to blinding sensors and disabling electronic systems. However, the development of truly effective laser weapons has been hampered by several limitations, including the size and power requirements of the laser systems, the efficiency of energy conversion, and the ability to precisely control the laser beam. Nanotechnology offers the potential to overcome these limitations.

One of the most significant challenges in laser weapon development is the efficient generation of high-power laser beams.

Current high-power laser systems often rely on bulky and inefficient components. Nano materials offer the possibility of creating more compact and efficient laser sources. For example, nano structured materials can be used to enhance the performance of laser gain media, the materials that amplify the laser light. Nano crystalline materials, with their high surface area and unique optical properties, can improve the absorption and emission of light, increasing the efficiency of the laser. Furthermore, nano materials can be used to create more efficient optical components such as mirrors, lenses, and waveguides, which are essential for directing and focusing the laser beam. Meta surfaces, artificial structures engineered at the nanoscale, can manipulate the properties of light with unprecedented precision, enabling the development of advanced optical components with superior performance. These advanced components can be significantly smaller and lighter than their conventional counterparts, leading to miniaturization of the entire laser system.

Another crucial aspect of laser weapon development is the control and precision of the laser beam. The ability to accurately target and focus the laser beam is critical for ensuring effective engagement of enemy assets. Nanotechnology can enhance the precision of laser systems through the development of advanced beam-steering mechanisms. Nano structured materials with precisely controlled optical properties can be used to create adaptive optical systems capable of compensating for atmospheric distortions, ensuring the accurate delivery of the laser beam over long distances. These systems can continuously monitor and adjust the shape of the wave front of the laser beam, correcting for turbulence and other atmospheric effects that can degrade the beam quality. Furthermore, nanomaterials can be used to create advanced sensors for monitoring the laser beam parameters and the target characteristics, providing real-time feedback for improving targeting accuracy.

In addition to high-powered lasers, electromagnetic pulse (EMP) weapons also represent a significant area where nanotechnology can have a substantial impact. EMP weapons generate intense bursts of electromagnetic radiation that can disable electronic systems over a wide area. These weapons can be particularly effective against electronic warfare systems, communications networks, and critical infrastructure. Conventional EMP weapons, however, are often large and unwieldy, making them difficult to deploy and operate.

Nanotechnology offers the potential for miniaturizing EMP weapons and enhancing their efficiency. Nano structured materials can be used to create smaller and more efficient EMP generators. For instance, nano materials with high electrical conductivity can be used to improve the performance of high-voltage switches and other critical components of the EMP generator. These switches are responsible for rapidly switching on and off the high-voltage current needed to generate the EMP burst. Faster switching speeds lead to shorter, more intense EMP pulses. Nano materials can also enhance the efficiency of energy storage devices used in EMP weapons. Advanced energy storage systems, such as super capacitors based on nano structured materials, can provide the high power pulses required for effective EMP generation in a compact form.

Furthermore, nanotechnology can be used to develop advanced materials for shielding against EMP attacks. Nano materials with unique electromagnetic properties can be used to create lightweight and effective shielding against EMP radiation. These materials can absorb or reflect the electromagnetic energy, protecting sensitive electronic systems from damage. The design and optimization of these shielding materials require advanced computational modelling and simulations at the nano scale, where the interaction of electromagnetic waves with nanostructures is highly complex.

The potential impact of miniaturized and more powerful DEWs on warfare is profound. More compact and efficient laser and EMP weapons could be integrated into a wider range of platforms, from aircraft and drones to ground vehicles and even handheld devices. This would allow for a more flexible and widespread deployment of DEWs, changing the nature of combat significantly. The increased power and precision of DEWs could lead to more effective and targeted engagements, reducing collateral damage. However, the development and deployment of such weapons also raise significant ethical and strategic considerations. The potential for misuse and escalation of conflict is a major concern. The international community needs to establish clear norms and regulations to govern the development and use of DEWs to ensure that these powerful technologies are used responsibly and in accordance with international law.

The exploration of advanced weaponry using nanotechnology naturally leads us to consider perhaps the most exotic and challenging concept: antimatter weapons. Antimatter, the elusive counterpart to ordinary matter, possesses the unique property of annihilating upon contact with its matter equivalent, releasing an immense amount of energy—far exceeding even nuclear reactions. This staggering energy density has captivated scientists and military strategists alike, powering the theoretical exploration of antimatter weapons for decades. However, the practical realization of such weapons faces formidable challenges, primarily centred around the creation, storage, and controlled release of antimatter. This is where nanotechnology steps in, offering potential solutions to these seemingly insurmountable obstacles.

The fundamental hurdle in developing antimatter weapons lies in the containment of antimatter. Antiprotons, for example, annihilate upon contact with even a single proton from ordinary matter. This means that a perfect vacuum is not sufficient; even a single residual gas molecule can trigger annihilation. Current containment methods rely primarily on electromagnetic fields, specifically Penning traps, to suspend antiprotons in a near-vacuum environment. These traps are sophisticated but bulky and energy-intensive, making them impractical for weapons. Nanotechnology, however, proposes novel approaches to improve containment significantly.

One promising avenue is the development of nano scale containment structures. Imagine structures composed of nan otubes or other nano materials meticulously engineered to create a near-perfect vacuum and to prevent any contact between antimatter and ordinary matter. These structures could potentially incorporate advanced materials with unique properties, such as extreme surface passivation to minimize interactions, or even materials exhibiting exotic quantum effects that could effectively shield the antimatter from its surroundings. Such structures would need to be incredibly robust, capable of withstanding the intense electromagnetic forces and the immense energy release upon accidental annihilation events. Moreover, the materials themselves must be radiation-hardened to resist the high-energy particles produced during annihilation.

Beyond simple containment, controlling the release of antimatter is equally crucial. For a weapon, the antimatter must be released precisely and at the desired moment to achieve the intended destructive effect. Nanotechnology could play a crucial role in designing mechanisms for controlled release. This could involve integrated nano scale valves or switches triggered by external stimuli, allowing for precise control over the rate and direction of antimatter ejection. Alternatively, sophisticated nano engineered structures could be designed to trigger a controlled annihilation cascade, carefully directing the resulting energy release.

The energy release itself poses another major challenge. The annihilation of even a small amount of antimatter generates a colossal amount of energy in a short burst. This necessitates the development of advanced energy management systems to harness and direct this energy effectively. Nanotechnology could help develop highly efficient energy-conversion mechanisms to transform the initial annihilation energy into other forms, such as directed energy beams (lasers or particle beams) or highly concentrated shockwaves. This controlled conversion is critical for maximizing the destructive potential while minimizing collateral damage. Sophisticated nanoscale systems could be deployed to rapidly disperse the resulting energy and radiation, further reducing the impact zone.

However, the challenges are far from trivial. The creation of antimatter is itself an energy-intensive process, requiring vast amounts of power far exceeding any current energy source suitable for weapons. Furthermore, the safety protocols involved in handling antimatter would need to be rigorous and fail-safe, minimizing the risk of accidental annihilation. The development of advanced sensors and safety systems at the nano scale is crucial for real-time monitoring and control.

Beyond the technological hurdles, ethical and strategic considerations are paramount. The immense destructive potential of antimatter weapons raises profound ethical concerns. The potential for accidental annihilation, even during testing phases, could have catastrophic consequences. The sheer destructive power of such weapons could significantly alter the balance of power on a global scale, potentially leading to an arms race with unimaginable consequences. The development and deployment of antimatter weapons would likely necessitate a radical reassessment of international security frameworks and norms.

Moreover, the cost of research, development, and deployment would be astronomical, diverting substantial resources from other crucial areas. The potential for misuse, accidental or intentional, necessitates careful consideration of the ethical, legal, and societal implications. The creation of stringent international protocols and regulations would be crucial to ensure responsible development and prevent the proliferation of this potentially devastating technology.

Therefore, while nanotechnology offers tantalizing possibilities for overcoming the significant technical challenges associated with antimatter weapons, the path to their practical realization remains exceptionally long and fraught with obstacles. Even if these technological hurdles are surmounted, the ethical, strategic, and economic considerations would demand a profound and ongoing global debate before any deployment could be even contemplated. The focus should remain on exploring the peaceful applications of nanotechnology and antimatter research, while simultaneously striving to prevent the misuse of these potentially revolutionary technologies. The future of antimatter weapons is, therefore, inextricably linked not only to scientific progress but also to the collective wisdom and responsible actions of humanity.

Nano thermites represent a significant advancement in energetic materials, offering a dramatic increase in reactivity and energy release compared to their conventional counterparts. These materials are based on the controlled mixing of nanoscale metal particles, typically aluminium (Al), with a metal oxide, such as iron oxide (Fe2O3) or copper oxide (CuO). The nano scale dimensions are crucial to their enhanced performance. In conventional thermites, the reactants are in micron-sized particles, limiting the surface area available for reaction. Nanothermites, however, possess vastly increased surface area due to the significantly smaller particle size, leading to a much faster and more vigorous reaction. This increased surface area promotes intimate mixing of reactants, accelerating the heat and mass transfer processes crucial for ignition and combustion.

The synthesis of nano thermites is a crucial aspect of their application. Several techniques are employed to achieve the desired nano scale mixing and particle size distribution. One common method is high-energy milling, where the metal and metal oxide powders are subjected to intense mechanical forces in a milling device. This process breaks down the particles to the nano scale while simultaneously promoting mixing. Other methods include sol-gel processing, where metal precursors are dissolved in a liquid and then undergo controlled chemical reactions to form a gel containing the desired metal and oxide particles. Spray pyrolysis is another technique that involves spraying a precursor solution into a hot furnace, leading to the formation of nano scale particles that can be subsequently mixed and processed.

Careful control of the particle size, morphology, and composition is crucial for optimizing the performance of nano thermites. The precise control over these parameters allows tailoring the reaction characteristics, including the burn rate, temperature, and overall energy output. This precise tailoring allows for their adaptation to a variety of applications. For example, a fast-burning nano thermite might be ideal for initiating a larger explosive event, while a slower-burning variant could be used in a sustained propulsion system.

The handling and safety aspects of nano thermites require special attention due to their high reactivity and potential for uncontrolled ignition. The fine particles are highly sensitive to friction, impact, and electrostatic discharge. Therefore, careful procedures are required during their synthesis, storage, and use. These precautions typically include the use of inert atmospheres, controlled humidity environments, and appropriate grounding equipment to minimize the risk of accidental ignition. Specialized protective clothing and equipment are also essential when handling these materials.

The applications of nano thermites extend to various fields, but their potential in advanced weaponry is particularly notable. In incendiary devices, nano thermites provide a significant improvement in the heat release and burn duration, resulting in more efficient and devastating fire effects. Their rapid combustion can be used to create highly localized intense heat pulses, allowing for the penetration and ignition of various materials, including armoured vehicles and fortifications. The high temperature produced can also be used to initiate or enhance other explosive reactions, leading to a cascade effect.

In shaped charges, nanothermites can significantly enhance the performance by increasing the jet velocity and penetration power. The intense heat and pressure generated during the reaction contribute to the formation of a high-velocity jet of molten metal, which can penetrate thick armor. The precise control over the burn rate allows optimization of the jet formation and its overall effectiveness.

Furthermore, nano thermites could find application in other advanced weapon systems. Their energy density and rapid combustion make them potential candidates for use in propellants, pyrotechnics, and even as components in more sophisticated weaponry under development. Research continues to explore the potential of nano thermites in diverse military applications, driving the innovation and development of even more powerful and precise weaponry.

However, the development and application of nano thermites are not without ethical and safety considerations. The enhanced reactivity and potential for unintended consequences require rigorous testing, safety protocols, and careful regulatory oversight. The potential for misuse of these highly energetic materials is a critical concern, necessitating international cooperation to prevent the proliferation of such technologies. Just as with antimatter, the balance between scientific advancement and responsible development is crucial. The pursuit of these potent materials must be coupled with a deep understanding of their implications and a commitment to responsible development and deployment. The focus must remain on maximizing the benefits while minimizing potential harm, ensuring that these powerful technologies are used responsibly for peaceful applications and not deployed in ways that might cause harm or destabilize the world. The long-term impacts of these materials must be carefully considered and studied, allowing for an informed approach to their implementation in diverse fields.

Plasma, the fourth state of matter, represents a fascinating frontier in weapons technology. Unlike conventional explosives or even directed energy weapons like lasers, plasma weapons leverage the unique properties of ionized gas to inflict damage. This ionized state, characterized by a significant number of free electrons and ions, allows for highly energetic and controllable interactions. The potential for focused, high-temperature streams of plasma capable of cutting through materials or disrupting electronics makes plasma weapons a subject of considerable interest in military circles and advanced defense research. However, realizing this potential requires sophisticated technology, and here, nanotechnology steps in to play a crucial enabling role.

One of the key challenges in developing effective plasma weapons is the efficient generation and sustained control of high-temperature plasma. Traditional methods often involve high-voltage discharges or powerful lasers, which can be bulky and energy-intensive. Nanotechnology offers a pathway to miniaturization and improved efficiency. For example, the use of nano structured electrodes in plasma generation devices can dramatically enhance the efficiency of electrical discharge, resulting in more powerful plasma streams with reduced energy consumption. Nano materials can also be employed to modify the surface properties of electrodes, increasing their resistance to erosion and prolonging their lifespan under the intense conditions of plasma generation.

The control and direction of plasma streams are equally crucial. Plasma, being a highly reactive and energetic substance, is inherently difficult to confine and direct accurately. This challenge is being addressed through the development of advanced plasma guiding and focusing systems that leverage nanotechnology. For example, nano structured materials can be used to create intricate magnetic field configurations for precise plasma confinement and steering. These systems might incorporate arrays of nano scale magnets or superconducting materials to manipulate the movement of charged particles in the plasma, effectively directing the energetic stream towards a target. The precision of these guiding systems is paramount in weapon applications, demanding intricate control over the plasma's trajectory. Nanotechnology enables the creation of miniature and precisely tuned components to achieve this accuracy.

The inherent instability of plasma is another major hurdle in weapon development. The tendency of plasma to dissipate quickly, particularly in open air, makes sustained, long-range operation challenging. Nano materials can address this issue by modifying the surrounding environment of the plasma stream. The incorporation of nano-engineered aerosols or coatings can alter the plasma's interaction with its surroundings, reducing its rate of cooling and dissipation, leading to a longer effective range and enhanced damage capabilities. The selection of these nano materials would be carefully calibrated to optimize their interaction with the generated plasma, thereby enhancing the weapon's overall effectiveness.

The nature of damage inflicted by plasma weapons can vary depending on the parameters of the plasma stream — temperature, density, and velocity. A high-temperature, high-velocity plasma stream can act as a highly effective cutting tool, capable of slicing through materials and leaving clean, precise cuts. The focused energy of the plasma can cause rapid heating and vaporization of the target material. The same plasma, if deployed with different parameters, could be designed to induce electromagnetic pulses (EMPs), thereby disrupting electronic systems. This versatility adds to the allure of plasma weapons as a potentially disruptive technology in the field of warfare.

The development of such weaponry also raises crucial safety considerations. The handling and control of high-energy plasma streams necessitate robust safety protocols to prevent accidents. Nanotechnology plays a role in enhancing these protocols. For instance, nano materials can be used in advanced sensor systems for detecting plasma leaks or malfunctions in the generating system, providing immediate feedback and triggering safety measures to prevent exposure to dangerous levels of radiation or high temperatures. Furthermore, nano structured barriers and containment systems could be employed to isolate the plasma and prevent any accidental release.

The material science challenges in creating effective plasma weapons are considerable. The intense temperatures and pressures involved demand materials with exceptional thermal and mechanical properties. This is where the potential of nano materials really shines. For example, carbon nano tubes, graphene, and other advanced nano materials can provide the necessary resilience and conductivity to withstand the extreme conditions within the plasma generation and control systems. The use of these materials permits a reduction in the size and weight of the weapon system, allowing for greater portability and deployment flexibility. Furthermore, nanotechnology allows for the creation of self-healing materials. These materials can automatically repair minor damage caused by the harsh plasma environment, extending the lifespan of the weapon system.

However, the path to practical plasma weapons is not without considerable obstacles. Aside from the challenges in generating, controlling, and directing plasma streams, the issue of power supply is crucial. Generating and sustaining high-power plasma streams demands considerable energy. While conventional power sources might suffice for some applications, the desire for mobile, portable plasma weapons requires significant advancements in energy storage and conversion technologies. Here again, nanotechnology potentially contributes through the development of high-energy density batteries and advanced energy-harvesting systems.

The economic and logistical implications of plasma weapons development and deployment are also significant. The production of nano materials and the construction of sophisticated plasma generation and control systems require advanced manufacturing techniques and substantial financial investment. The cost-effectiveness and scalability of plasma weapons compared to other weapon systems are therefore critical considerations. Research into sustainable and cost-effective production methods for the necessary nano materials will be crucial in determining the feasibility and widespread deployment of plasma weapons.

The development of advanced weaponry hinges not only on the destructive power of the weapon itself but also on its ability to withstand the rigors of deployment and combat. This necessitates the creation of weapon systems that are exceptionally durable and resistant to damage. Nanotechnology offers a transformative path towards achieving this goal, enabling the creation of weapon casings and components with unprecedented levels of hardness, resilience, and protection.

One crucial area where nano materials excel is in enhancing the protective casings of weapons. Conventional materials, even high-strength alloys like steel and titanium, are susceptible to damage from impacts, extreme temperatures, and corrosive environments. Nano materials, however, offer superior strength-to-weight ratios and enhanced resistance to various forms of degradation. For example, carbon nano tubes (CNTs) exhibit exceptional tensile strength, far exceeding that of steel, while possessing a significantly lower density. Incorporating CNTs into composite materials used for weapon casings can drastically improve their ability to withstand impacts and ballistic damage. The precise arrangement and alignment of CNTs within the composite matrix can be tailored to optimize the material's response to specific types of stress, maximizing its protective capabilities.

Graphene, a single layer of carbon atoms arranged in a hexagonal lattice, is another remarkable nano material with exceptional mechanical and protective properties. Its exceptional strength, flexibility, and high thermal conductivity make it ideal for applications requiring both high strength and the ability to dissipate heat efficiently. Graphene can be incorporated into composite materials for weapon casings to enhance their resistance to both mechanical and thermal stress. Moreover, graphene's unique electrical properties can be leveraged to create self-monitoring protective layers capable of detecting and reporting damage in real time, providing valuable feedback for maintenance and repair.

Beyond carbon-based nano materials, other promising candidates for weapon hardening include boron nitride nano tubes (BNNTs) and ceramic nano particles. BNNTs possess exceptional thermal stability and chemical inertness, making them highly resistant to high-temperature environments and corrosive substances. This characteristic is particularly valuable for weapons deployed in extreme conditions, such as high-altitude or underwater applications. Ceramic nano particles, such as silicon carbide (SiC) and aluminium oxide (Al2O3), can significantly enhance the hardness and wear resistance of composite materials, providing exceptional protection against abrasion and erosion. These nano particles can be dispersed within a polymeric matrix, creating a composite material with enhanced durability and impact resistance, suitable for weapon components exposed to significant mechanical stress.

The integration of nano materials into weapon components extends beyond just the casings. Internal components, such as gears, bearings, and electronic circuitry, can also benefit from nano material enhancements. For example, nano sized lubricants can dramatically reduce friction and wear in moving parts, enhancing efficiency and lifespan. Nano-coatings can provide enhanced corrosion resistance and improve the reliability of electronic systems subjected to harsh operating conditions. This approach not only improves the durability of individual components but also enhances the overall functionality and reliability of the weapon system as a whole.

The effectiveness of nano material-enhanced weapon systems depends significantly on the specific type of threat they are designed to withstand. For example, weapons intended for use in high-impact environments, such as those encountered in close-quarters combat, would benefit from materials with high impact resistance. This might involve using composites reinforced with CNTs or ceramic nano particles, or employing specialized designs to distribute and absorb impact energy effectively. On the other hand, weapons designed for use in environments with high levels of radiation or extreme temperatures would require materials with enhanced radiation resistance and thermal stability, such as BNNTs or specialized ceramic composites.

The development of advanced armour materials is another critical application of nanotechnology in weapon systems. Traditional armor materials, such as steel and Kevlar, are continually being improved, but nanomaterials offer the potential for significant advancements in this area. Nano materials like graphene and CNTs can be integrated into composite armour structures, resulting in significantly enhanced ballistic resistance and protection against penetration. The layered structure of these composite materials, with alternating layers of high-strength nano materials and more flexible polymers, allows them to absorb and distribute impact energy more effectively than conventional armor, thereby minimizing damage to the wearer.

Moreover, the incorporation of shape-memory alloys at the nano scale level offers potential for self-healing armour. These materials can remember their original shape and return to it after deformation, effectively repairing minor damage caused by impacts. This self-healing capability can extend the lifespan of armour systems and provide enhanced protection for extended periods of use. The development of these materials necessitates advanced fabrication techniques such as chemical vapour deposition and solution-based processing to ensure precise control over the nano scale architecture of the composite structures, maximizing their effectiveness.

However, the widespread adoption of nano materials in weapon systems is not without its challenges. One significant hurdle is the cost of production. Many nano materials, particularly those with exceptional properties, are currently expensive to produce on a large scale. This high cost can limit their widespread implementation in military applications, particularly where cost-effectiveness is a primary concern. Research into more efficient and cost-effective synthesis methods is crucial for the broader adoption of nanomaterials in weapon hardening and protection technologies.

Another challenge lies in the long-term reliability and durability of nano material-based composites. While nano materials offer significant improvements in strength and resilience, their long-term behaviour under extreme conditions, such as repeated exposure to high temperatures or corrosive environments, needs to be thoroughly characterized. Extensive testing and validation are necessary to ensure that the performance of nano material-enhanced weapon systems remains consistent and reliable throughout their intended lifespan. This requires sophisticated characterization techniques, including advanced microscopy, spectroscopy, and mechanical testing, to monitor the long-term stability and performance of these materials under various conditions. The development of predictive models capable of assessing the long-term durability and reliability of these materials under operational conditions would also be instrumental.

Furthermore, the potential environmental impact of nano materials needs to be carefully considered. The lifecycle assessment of nano material production, use, and disposal must be thoroughly evaluated to mitigate any potential risks to the environment and human health. This necessitates careful management of the production processes to minimize the release of nano particles into the environment, as well as the development of sustainable disposal methods for spent materials. Responsible development and deployment of nano material-based weapon systems require strict adherence to environmental protection regulations and a commitment to minimizing any negative impact on the environment.

The lethality and effectiveness of any weapon system are fundamentally determined not only by its robustness but also by the precision and sophistication of its guidance and control mechanisms. This is where nanotechnology's transformative potential truly shines, offering unprecedented opportunities to revolutionize weapon accuracy, responsiveness, and battlefield awareness. The integration of nano scale materials and devices promises a new generation of weapon systems with capabilities far exceeding those of their conventional counterparts.

One of the most impactful applications of nanotechnology in weapon guidance is the development of highly sensitive and precise nano sensors. These minuscule devices, often measuring just nano meters in size, can detect a wide range of signals and environmental parameters with extraordinary accuracy. In the context of weapon guidance, nano sensors can be integrated into the guidance system to provide real-time feedback on the weapon's trajectory, surrounding environment, and the target's characteristics. For example, nano sensors could detect subtle changes in atmospheric conditions, such as wind speed and direction, allowing for real-time corrections to the weapon's trajectory and ensuring greater accuracy. This is particularly crucial for long-range weapons where even minor deviations in atmospheric conditions can significantly impact accuracy.

Furthermore, nano sensors can provide critical information about the target itself. Imagine a guided missile equipped with nano sensors capable of detecting the target's heat signature, electromagnetic emissions, or even subtle chemical changes in the surrounding environment. This information can be used to refine the targeting algorithms and ensure that the weapon engages the intended target with maximum effectiveness, minimizing collateral damage. For example, nano sensors could distinguish between a legitimate military target and nearby civilian structures, dramatically reducing the risk of civilian casualties. Such discrimination capabilities are difficult, if not impossible, to achieve with conventional sensing technologies.

Another critical aspect of weapon guidance systems is the ability to react quickly and precisely to changes in the target's position or the surrounding environment. This is where nano actuators play a crucial role. Nano actuators are microscopic devices that can produce mechanical movement or control the flow of energy or fluids at the nano scale. In weapon guidance systems, nano actuators can be used to rapidly adjust the weapon's trajectory or control the deployment of countermeasures in response to unexpected events. The speed and precision of nan oactuators far exceed those of conventional actuators, allowing for a level of responsiveness that was previously unattainable.

Consider a guided missile approaching a target that suddenly employs countermeasures such as chaff or flares. A conventional guidance system might struggle to distinguish between the legitimate target and the decoys, potentially leading to a failed mission. However, a guidance system equipped with nano actuators could quickly and precisely adjust the missile's trajectory, avoiding the decoys and engaging the target with precision. This level of responsiveness significantly increases the chances of a successful mission outcome and improves overall battlefield effectiveness.

The integration of nanotechnology also allows for miniaturization of the guidance and control systems. This is crucial for applications where size and weight are critical constraints, such as in guided munitions or unmanned aerial vehicles. The ability to create smaller, lighter, and more energy-efficient guidance systems opens up new possibilities for weapon design and deployment. For instance, it could enable the development of extremely precise, yet compact, weapons that can be easily integrated into existing platforms, expanding their operational capabilities.

Beyond improving the accuracy and responsiveness of individual weapons, nanotechnology has the potential to revolutionize battlefield situational awareness. Networks of nano sensors deployed across a wide area can provide real-time information on enemy movements, troop concentrations, and other critical battlefield parameters. This information can be used to improve tactical decision-making, optimize force deployment, and enhance overall combat effectiveness. Nano sensors could be disseminated through various methods, such as being integrated into the environment (e.g., embedded in the ground or dispersed in the air), providing a persistent and distributed sensing capability.

The information gathered by these networks of nano sensors can be processed and analyzed using advanced algorithms to create a comprehensive picture of the battlefield. This improved situational awareness allows commanders to anticipate enemy actions, pre-emptively deploy forces, and minimize potential risks to friendly personnel. The level of detail and real-time data provided by nano sensor networks could drastically improve operational decision making, offering a significant advantage in modern warfare.

However, the implementation of nanotechnology in weapon guidance and control systems faces challenges. The cost of manufacturing nano sensors and nano actuators can be relatively high, limiting their widespread adoption. Furthermore, the integration of these nano scale devices into larger weapon systems requires advanced manufacturing techniques and careful consideration of compatibility and reliability. The long-term durability and performance of nano sensors in harsh battlefield environments also need to be thoroughly evaluated. The complexity of data transmission and processing from large networks of nano sensors pose additional challenges.

Moreover, the potential for misuse of such technology raises ethical concerns. The development of highly accurate and responsive weapons systems requires careful consideration of the implications for international security and the potential for escalation of conflict. Robust safeguards and international agreements will be crucial to prevent the proliferation of these technologies and ensure responsible development and deployment.

Despite these challenges, the integration of nanotechnology into weapon guidance and control systems holds immense potential. It promises a future where weapons are not only more durable and resilient but also significantly more accurate, responsive, and adaptable. The advancements in nano sensors, nano actuators, and miniaturized systems will continue to drive innovation in weapon design, leading to a transformation in battlefield capabilities. The ethical considerations surrounding the development and use of this technology must however remain at the forefront of any research and development efforts. This requires careful consideration and international cooperation to ensure responsible innovation and minimize the risk of unintended consequences. The future of warfare will undoubtedly be shaped by the advancement and application of nanotechnology, making careful planning and strategic decision-making crucial to manage this rapidly evolving technology. Continuous research and development, coupled with a strong ethical framework, will be vital to realize the full potential of nanotechnology in weapon guidance and control systems while mitigating potential risks.

The advent of nanotechnology-enhanced weapons systems presents a profound challenge and opportunity for military doctrine and strategy. The unique characteristics of these weapons – their potential for miniaturization, increased precision, enhanced lethality, and novel functionalities – necessitate a reassessment of existing operational paradigms and a proactive adaptation to the evolving battlefield landscape.

One of the most immediate implications lies in the potential transformation of battlefield tactics. The precision offered by nanotechnology-enabled munitions, for example, could drastically reduce collateral damage compared to conventional weaponry. This increased precision could allow for more surgical strikes, targeting specific enemy assets or personnel with minimal impact on civilian populations or infrastructure. However, this very precision also raises ethical considerations, particularly concerning the potential for preemptive strikes and the blurring of lines between combatants and non-combatants in complex operational environments. Furthermore, the miniaturization afforded by nanotechnology could lead to the development of smaller, more easily concealable weapons, making them more accessible to non-state actors and potentially destabilizing regional security.

The development of nanotechnology-based directed energy weapons, such as high-powered lasers or electromagnetic pulse (EMP) weapons powered by miniaturized nuclear fission or fusion reactors, could revolutionize naval and air warfare. These weapons offer the potential for instantaneous engagement at long ranges, potentially negating the advantages of traditional kinetic weapons. The strategic implications are significant, potentially altering the balance of power and necessitating the development of new defensive technologies and strategies. The vulnerability of existing electronic systems to EMP attacks also raises crucial concerns about the resilience of military infrastructure and command-and-control systems. The integration of nano materials into these systems to enhance their resilience to EMP effects is a key area of ongoing research and development.

The potential for asymmetric warfare is significantly amplified by the accessibility of nanotechnology-enhanced weapons. The miniaturization capabilities of nanotechnology could allow smaller, less technologically advanced nations or non-state actors to develop and deploy relatively sophisticated weaponry, potentially upsetting the existing balance of power. This necessitates a reassessment of military strategies and resource allocation to counter the threat posed by these unconventional actors. The implications for counter-terrorism strategies are particularly relevant, requiring a focus on disrupting the supply chains and networks responsible for procuring or developing these advanced weapons.

Arms control and non-proliferation efforts face new challenges in the age of nanotechnology. The potential for the development of more powerful and easily concealed weapons of mass destruction, including nuclear weapons enhanced by nano materials, necessitates a re-evaluation of existing international treaties and agreements. The verification of compliance with arms control agreements becomes significantly more complex when dealing with miniaturized and potentially undetectable weapons. This necessitates the development of new verification technologies and methods, potentially leveraging advanced analytical techniques to identify the presence of nano materials in suspect weapons or components. International cooperation and the establishment of robust verification protocols are vital to prevent the proliferation of these advanced weapons.

The rapid pace of technological advancement in nanotechnology demands an adaptable and agile military doctrine. The development of future-proof strategies that can respond to the rapid evolution of nanotechnology-enhanced weapons is crucial. This includes investing in research and development efforts to develop countermeasures and defensive technologies, coupled with advanced simulations and war gaming exercises to anticipate and respond to evolving threats. The development of flexible and modular weapons systems, adaptable to different threat scenarios, is also critical. Regularly updating military training programs and doctrine to reflect the capabilities and limitations of nanotechnology-enhanced weapons is essential to maintain operational effectiveness.

Furthermore, the potential impact on intelligence gathering and analysis must be considered. The development of nano scale sensors and surveillance technologies could revolutionize reconnaissance and target acquisition. However, this also raises concerns about privacy and potential abuse of these technologies. The ethical implications of using nanotechnology for surveillance and information gathering must be carefully considered. The establishment of clear ethical guidelines and oversight mechanisms is crucial to ensure responsible use of these advanced technologies.

Beyond the immediate battlefield implications, the long-term impact on geopolitical stability must be considered. The potential for power shifts and the emergence of new global players due to advancements in nanotechnology-enhanced weapons requires careful analysis and strategic foresight. The ability of smaller nations or non-state actors to acquire advanced weapons capabilities could destabilize existing regional security dynamics and lead to unexpected conflicts. Proactive diplomatic efforts and international collaborations are necessary to prevent the escalation of tensions and to foster a stable global security environment.

Future Trends and Technological Advancements

The rapid pace of technological advancement necessitates a look toward the future, specifically examining the potential enhancements in AI and autonomy that will fundamentally reshape the capabilities and applications of nano drones. This section explores these anticipated advancements, focusing on improvements in decision-making, swarm coordination, and adaptive behaviors, while acknowledging the considerable challenges inherent in creating truly intelligent and adaptable autonomous systems.

One of the most significant areas of future development lies in enhancing the decision-making capabilities of nano drones. Currently, many nano drones operate based on pre-programmed instructions or rely on remote human control. However, future generations will likely incorporate more sophisticated AI algorithms allowing for greater autonomy in complex environments. This involves moving beyond simple reactive behaviors to more proactive and anticipatory decision-making. Machine learning, particularly deep learning techniques, offers a promising pathway to this enhanced autonomy. Deep learning algorithms, trained on vast datasets of sensor information and environmental conditions, can learn to identify patterns, predict outcomes, and make decisions with minimal human intervention. This would enable nano drones to navigate unpredictable terrains, adapt to changing environmental conditions, and even exhibit a degree of strategic thinking. For instance, a swarm of nano drones tasked with surveillance could utilize deep learning to optimize their search patterns, dynamically adjusting their positions based on the changing terrain and the detection of potential targets.

The development of more sophisticated AI algorithms also promises to significantly improve swarm coordination. Currently, coordinating large swarms of nano drones requires considerable computational power and intricate communication protocols. However, future advancements in distributed AI and swarm intelligence could enable swarms to operate with greater autonomy and efficiency. Decentralized algorithms, where individual drones make decisions based on local information and communicate only with nearby drones, could drastically reduce the computational burden and improve the resilience of the swarm. This distributed approach would make the swarm more robust to interference or the loss of individual drones. Future swarms could exhibit emergent behaviours, where complex coordinated actions arise from simple interactions between individual drones, akin to the behavior of ant colonies or bird flocks. Such emergent behaviours could dramatically enhance the effectiveness of swarms in tasks such as coordinated surveillance, search and rescue operations, or even limited combat scenarios.

Adaptive behaviours are another crucial area for future development. The ability of nano drones to adapt to changing conditions and unexpected events is critical for their successful deployment in diverse and challenging environments. This requires AI algorithms that can learn from experience, adapt to new situations, and adjust their strategies accordingly. Reinforcement learning, a type of machine learning where an agent learns to optimize its actions through trial and error, offers a powerful tool for developing adaptive behaviors in nano drones. For instance, a nano drone tasked with navigating a cluttered environment could use reinforcement learning to learn the optimal path through the obstacles, adapting its strategy based on the specific challenges encountered. Such adaptive behaviors are crucial for ensuring the robustness and reliability of nano drones in unpredictable scenarios.

Beyond these specific advancements, the integration of more powerful onboard sensors and improved communication technologies will further enhance the autonomy and capabilities of nano drones. Higher-resolution cameras, advanced sensors for detecting chemical or biological agents, and improved communication protocols will provide the drones with more detailed information about their environment, enabling them to make more informed decisions. The development of novel energy harvesting techniques, such as utilizing solar or vibrational energy, will also extend their operational endurance, further enhancing their autonomy and operational range.

However, the development of truly intelligent and adaptable autonomous systems presents significant challenges. One major hurdle is the need for robust and reliable AI algorithms that can handle uncertainty and unexpected events. The unpredictable nature of real-world environments requires AI systems that can cope with noise, incomplete information, and unforeseen circumstances. This necessitates the development of AI algorithms that are not only powerful but also robust and reliable. Furthermore, ensuring the safety and security of autonomous nano drones is paramount. The potential for malicious use or unintended consequences requires the implementation of robust safety mechanisms and security protocols to prevent accidents or misuse. This includes the development of fail-safe mechanisms, cyber security measures to prevent hacking, and ethical guidelines to ensure responsible development and deployment.

Speculative discussions about future AI capabilities often involve the integration of advanced cognitive architectures. These architectures would enable nano drones to exhibit more human-like reasoning and decision-making abilities, including the capacity for problem-solving, planning, and learning from experience. Such advanced AI systems could revolutionize the applications of nano drones, enabling them to perform increasingly complex tasks with minimal human intervention. However, the development of such sophisticated AI systems poses significant technical and ethical challenges. The potential for unintended consequences and the ethical implications of highly autonomous systems demand careful consideration and proactive mitigation strategies.

The potential impact of these advancements on military applications is considerable. Enhanced AI and autonomy could transform nano drones into highly effective tools for surveillance, reconnaissance, and even limited combat operations. Swarms of highly autonomous nano drones could overwhelm enemy defenses, conduct precision strikes, or provide unparalleled situational awareness. However, the ethical implications of deploying autonomous weapons systems require careful consideration. The potential for unintended casualties and the lack of human control raise serious ethical concerns. International agreements and regulatory frameworks will be crucial to prevent an arms race in autonomous nano drone weaponry.

In the civilian sphere, the advancements in AI and autonomy could lead to widespread adoption of nano drones in various sectors. Enhanced autonomous capabilities will enable the use of nano drones for tasks such as environmental monitoring, infrastructure inspection, precision agriculture, and search and rescue operations. The ability of swarms to coordinate their actions and adapt to changing conditions will dramatically improve the efficiency and effectiveness of these applications. However, concerns about privacy and data security must be addressed to ensure the responsible deployment of this technology. Robust regulations and ethical guidelines will be crucial to prevent misuse and protect individual rights.

One of the most significant challenges in nano drone development is the inherent trade-off between size, weight, and performance. Smaller drones inherently possess less payload capacity, limiting battery life and the range of sensors and actuators they can carry. However, advancements in materials science are mitigating this limitation. The development of carbon nano tubes, graphene, and other advanced composites offers the potential for incredibly lightweight yet incredibly strong structural components. These materials boast significantly higher strength-to-weight ratios compared to traditional materials like aluminium or plastics, enabling the construction of smaller drones without sacrificing structural integrity. Moreover, their flexibility allows for the creation of drones with articulated wings or other dynamic components, enhancing agility and manoeuvrability.

Beyond structural components, advancements in battery technology are critical. The energy density of current battery technologies limits the flight time of nano drones. The pursuit of higher energy density batteries is an active area of research, with significant progress being made in solid-state batteries, lithium-sulfur batteries, and other promising technologies. These advancements promise a substantial increase in flight time, expanding the operational range and capabilities of nano drones. Furthermore, research into energy harvesting techniques, such as using solar energy or vibrational energy, is also gaining traction. Integrating miniature solar cells or piezoelectric generators could provide supplementary power sources, extending flight times even further.

The development of new materials also extends to the creation of more robust and resilient nano drones. The ability of a nano drone to withstand impacts, harsh weather conditions, and other environmental stressors is crucial for its operational reliability. Research into self-healing materials and coatings provides a promising pathway to enhance the durability of nano drones. These materials can repair minor damage autonomously, extending the operational lifespan and reducing maintenance requirements. This is particularly important for applications where drones are deployed in challenging or unpredictable environments.

Micro-manufacturing techniques play a pivotal role in realizing the potential of advanced materials. The creation of increasingly smaller and more complex components requires sophisticated fabrication processes. Advances in 3D printing, laser micromachining, and other precision manufacturing techniques enable the production of intricate structures and components at the micro scale. These techniques allow for the integration of complex sensor systems, actuators, and control electronics within extremely small packages. For instance, 3D printing allows for the creation of complex internal structures that optimize weight distribution and structural strength, maximizing payload capacity while minimizing weight. Laser micromachining enables the creation of extremely precise components, such as miniature gears, actuators, and optical systems, essential for the operation of advanced nano drones.

The development of micro-electromechanical systems (MEMS) is another crucial area of advancement. MEMS technology enables the creation of miniature sensors and actuators that are essential for the operation of nano drones. These miniature sensors can provide information about the drone's position, orientation, altitude, velocity, and environmental conditions. MEMS actuators can control the movement of the drone's rotors, wings, or other components. The miniaturization and integration of these MEMS components are crucial for reducing the size and weight of nano drones while maintaining their functionality.

The impact of these advancements on nano drone performance is substantial. Smaller, lighter drones with longer flight times and enhanced durability can be deployed in previously inaccessible areas.

For military applications, this translates to enhanced surveillance capabilities, improved reconnaissance missions, and more effective targeting systems. In civilian applications, these advancements enable the widespread adoption of nano drones for tasks such as environmental monitoring, infrastructure inspection, search and rescue operations, and precision agriculture. The ability to deploy swarms of small, agile, and long-endurance drones opens up a wide range of possibilities.

However, the miniaturization and materials science challenges remain significant. The manufacturing of extremely small components requires precise control over materials properties and manufacturing processes. The integration of complex electronic systems within extremely small packages poses significant challenges in terms of heat dissipation and power management. Furthermore, the cost of producing nano drones with these advanced materials and manufacturing techniques can be high, potentially limiting their widespread adoption.

Ongoing research efforts are addressing these challenges. Scientists and engineers are continuously exploring new materials with enhanced properties, developing more efficient energy storage solutions, and refining micro-manufacturing techniques. The development of self-assembly techniques, where components automatically assemble themselves, holds the promise of simplifying the manufacturing process and reducing costs. Furthermore, advancements in artificial intelligence and machine learning are playing a critical role in optimizing the design and performance of nano drones, enabling the creation of more efficient and robust systems.

The future of nano drone technology depends on the continued advancements in miniaturization and materials science. As new materials with superior properties are developed and more sophisticated manufacturing techniques are implemented, the capabilities of nano drones will continue to improve. Smaller, faster, more agile, and more durable drones will enable new applications across a wide range of sectors, transforming how we interact with the world. However, alongside these technological advancements, careful consideration must be given to the ethical implications and potential misuse of this technology, ensuring responsible development and deployment. The path towards truly revolutionary nano drone capabilities is paved by ongoing innovation in both the software and hardware aspects, demanding persistent research and development across multiple disciplines. The convergence of advanced materials, precise manufacturing, and sophisticated AI is shaping a future where these tiny machines will play an increasingly significant role. The implications, both positive and potentially negative, demand careful consideration and proactive planning by researchers, policymakers, and society as a whole.

The limitations imposed by current battery technology significantly constrain the operational capabilities of nano drones. Flight times are often short, restricting their range and the complexity of missions they can undertake. To overcome this, researchers are actively pursuing advancements in several key areas, focusing on both enhancing existing battery technology and exploring alternative energy harvesting methods.

One of the most promising avenues lies in the development of high-energy-density batteries. Lithium-ion batteries, while currently dominant, are reaching their theoretical limits in terms of energy density. Solid-state batteries represent a significant leap forward. Replacing the liquid electrolyte in conventional lithium-ion batteries with a solid electrolyte offers several advantages. Solid-state batteries are inherently safer, less prone to thermal runaway and potentially offer higher energy densities. This increased energy density translates directly into longer flight times for nano drones, expanding their operational capabilities. However, challenges remain in terms of manufacturing cost and scalability. The production of high-quality solid-state electrolytes remains a significant hurdle, but ongoing research is yielding promising results, and the potential payoff is substantial.

Beyond solid-state batteries, research is also focused on other battery chemistries, such as lithium-sulphur batteries. These batteries offer the potential for even higher energy densities than lithium-ion batteries, owing to the high theoretical capacity of sulphur. However, the practical implementation of lithium-sulphur batteries faces challenges, primarily concerning the cycling life and the tendency for sulphur to dissolve in the electrolyte. Despite these challenges, ongoing research and development efforts are addressing these limitations, bringing the promise of significantly extended flight times closer to reality.

Improvements in charging methods are also essential for maximizing the operational efficiency of nano drones. Wireless charging technologies are gaining traction, eliminating the need for cumbersome wired connections. Inductive charging, utilizing electromagnetic fields to transfer energy, is already being implemented in some commercial drones, although scaling this technology down for nano drones presents considerable challenges in terms of efficiency and range. However, advancements in near-field communication and resonant inductive coupling techniques are showing promise for highly efficient wireless power transfer at the micro scale, paving the way for seamless charging of nano drone swarms.

Beyond battery technology, the exploration of alternative energy harvesting methods offers the potential for truly transformative advancements. Solar energy, for instance, presents an attractive option. The integration of miniature solar cells onto the drone's surface can provide a supplementary power source, extending flight time and operational endurance. Advancements in thin-film solar cell technology are crucial here, allowing for the development of lightweight, flexible solar cells that can be seamlessly integrated onto the drone's surface without compromising its aerodynamics. The efficiency of these miniature solar cells is a key consideration; maximizing their power output while minimizing their weight and size is a continuous challenge. However, the potential for near-continuous operation in sunlight-rich environments makes this an extremely compelling avenue of research.

Piezoelectric energy harvesting offers another promising alternative. Piezoelectric materials generate electricity in response to mechanical stress or vibration. This principle could be leveraged by integrating piezoelectric generators into the drone's structure. The vibrations generated during flight could then be harnessed to generate supplementary power, extending the drone's operational time. The efficiency of this method depends heavily on the intensity and frequency of vibrations, making its implementation challenging. Further research is needed to optimize the design and integration of piezoelectric generators to maximize their energy output and minimize their weight and size. Successful integration of this technology could provide a self-sustaining power source, particularly suitable for drones operating in environments with constant vibrations.

Fuel cells offer another avenue, although they present unique challenges at the nano scale. Miniaturizing fuel cells to a size compatible with nano drones requires significant advancements in materials science and micro-manufacturing techniques. The challenges include managing the flow of reactants and products within the confined space, optimizing the catalytic reactions, and ensuring efficient heat dissipation. While technically feasible, the complexities and limitations associated with miniaturizing fuel cells make it a less immediate solution compared to battery and energy harvesting improvements. However, their potential to offer significantly extended flight times makes ongoing research vital.

The choice of energy source ultimately depends on the specific application and operational environment of the nano drone. For short-duration missions in controlled environments, high-energy-density batteries may suffice. For longer-duration missions or those operating in environments with abundant sunlight, a hybrid approach combining batteries and solar energy harvesting may be the most effective. For drones operating in environments with constant vibrations, piezoelectric energy harvesting could prove particularly beneficial. The optimal energy solution will likely be application-specific, requiring a careful assessment of the mission requirements and environmental conditions.

Furthermore, the efficient management of energy is crucial for maximizing operational time. Power management systems, incorporating intelligent power switching and energy-efficient components, are essential for optimizing energy consumption. Advancements in low-power electronics and microprocessors are critical in this regard, allowing for the operation of sophisticated sensor systems and onboard processing without excessive power drain. The development of sophisticated algorithms for autonomous power management will also be crucial, dynamically adjusting power consumption based on the mission requirements and available energy resources.

The development of efficient and reliable energy sources for nano drones is a critical aspect of their continued advancement. The synergy between improved battery technologies, alternative energy harvesting methods, and intelligent power management systems will unlock the full potential of these miniature autonomous vehicles. The ongoing research and development in this field are crucial for expanding their operational capabilities, extending their flight times, and ultimately enabling them to undertake more complex and challenging missions. The future of nano drones hinges on the successful integration of these advanced energy technologies, enabling their widespread adoption across diverse civilian and military applications. Continuous innovation in this area will ultimately define the boundaries of their capabilities and potential impact on future technological landscapes. The development and refinement of these power sources are not merely technological challenges; they are fundamental to achieving the full potential of these groundbreaking devices. The convergence of advancements in materials science, energy harvesting, and power management systems will dictate the extent to which nano drones will revolutionize various industries.

The miniaturization of sensor technology is crucial to the advancement of nano drone capabilities. While enhanced power sources extend operational time, equally vital is the ability to collect and process meaningful data efficiently within those extended operational windows. Current limitations in sensor size and power consumption are being aggressively addressed, leading to remarkable improvements in data acquisition and processing.

One of the most significant advancements lies in camera technology. High-resolution cameras, once the domain of larger, heavier drones, are now being successfully miniaturized for integration into nano drones. This miniaturization is driven by advancements in micro electromechanical systems (MEMS) technology, allowing for the fabrication of sophisticated imaging sensors on a microscopic scale. These miniature cameras are capable of capturing high-resolution images and video, providing significantly improved situational awareness. The development of specialized lenses, incorporating advanced optical designs, further enhances image quality, compensating for the limitations imposed by the small form factor.

Beyond simply increasing resolution, significant strides have been made in improving the image processing capabilities of nano drones. Advanced algorithms, leveraging machine learning and artificial intelligence, are being incorporated into on-board processing units to enhance image quality and extract relevant information from captured images and video. These algorithms can automatically adjust for lighting conditions, reduce noise, and enhance contrast, resulting in clearer and more usable images. Furthermore, advanced image processing techniques, such as image stitching and panoramic imaging, enable the creation of comprehensive visual representations of the environment, even from a limited vantage point.

The incorporation of advanced image recognition software transforms raw visual data into actionable intelligence. These algorithms can identify and classify objects, individuals, and even specific features within the captured imagery. This capability enables the nano drone to autonomously recognize and track targets, detect anomalies, and provide real-time situational awareness to operators. The continuous improvement of these algorithms, through machine learning and the use of large datasets, is leading to increasingly accurate and reliable target recognition capabilities. Real-time object identification is a significant leap forward, transforming the nano drone from a simple data-gathering device into a powerful intelligence-gathering asset.

Beyond visual sensing, the integration of other sensor modalities is greatly expanding the capabilities of nano drones. Chemical sensors, for example, can detect the presence of specific gases or volatile organic compounds (VOCs). This capability is particularly useful in environmental monitoring, search and rescue operations, and even counter-terrorism efforts. Miniaturized chemical sensors are becoming increasingly sophisticated, capable of detecting a wide range of compounds with high sensitivity and specificity.

Biological sensors are another area of rapid development. These sensors can detect the presence of biological agents, such as bacteria, viruses, or toxins. Such sensors have significant implications for medical applications, public health monitoring, and countermeasures against biological weapons. The challenge in this area is the miniaturization of highly sensitive biosensors, maintaining their accuracy and reliability in a small and rugged package. Advances in nanotechnology and micro fluidics are key to overcoming these challenges and enabling the integration of sophisticated biosensors into nano drones.

Environmental sensors, including temperature, humidity, pressure, and radiation sensors, provide comprehensive environmental data. This data is crucial for a range of applications, including weather forecasting, environmental monitoring, and search and rescue operations in challenging environments. These sensors are increasingly miniaturized and consume less power, making them suitable for integration into nano drones. The data they provide can be combined with other sensor data to create a more comprehensive understanding of the environment.

Data processing is a crucial aspect of the overall performance of nano drones. Even the most advanced sensors are of limited use if their data cannot be processed and transmitted efficiently. The processing power of on-board computers in nano drones is continuously increasing, allowing for more sophisticated data analysis and autonomous decision-making. The development of low-power, high-performance processors is key to this advancement.

The efficient transmission of collected data is equally important. Advancements in communication technologies, such as low-power wide-area networks (LPWANs) and other advanced wireless communication protocols, are crucial for enabling reliable and low-latency data transmission from nano drones. These protocols are optimized for low-power consumption, enabling long-range communication with minimal power drain. The challenge is maintaining a robust connection even in challenging environments with limited signal strength or interference. Developing resilient communication protocols that can overcome such challenges is key to unlocking the full potential of nano drones.

Data compression techniques are also essential for efficient data transmission. Reducing the size of data files allows for faster transmission speeds and reduces the energy required for transmission. Advanced compression algorithms, leveraging machine learning and artificial intelligence, are improving the efficiency of data compression. By only transmitting the most relevant data, energy efficiency is maximized, extending the operational lifespan of the nano drone. This selective data transmission focuses the energy budget on information that truly matters, enhancing the overall performance of the system.

The integration of advanced sensor technologies and efficient data processing capabilities marks a significant leap forward in the capabilities of nano drones. These advancements are transforming them from simple data-gathering devices into powerful autonomous systems capable of performing complex tasks and providing crucial real-time information. The ongoing research and development efforts in this field are crucial for unlocking the full potential of nano drones in various sectors, including military applications, environmental monitoring, and search and rescue operations. The synergy between improved sensor technology, advanced data processing algorithms, and efficient communication protocols will define the future of this rapidly evolving technology.

Furthermore, the development of robust data fusion techniques is crucial for integrating data from multiple sensors. Data from different sensor modalities can be combined to create a more complete and accurate picture of the environment. For example, combining visual data from a camera with chemical data from a gas sensor can provide a more comprehensive understanding of a hazardous situation. Advanced algorithms are being developed to seamlessly integrate this diverse data, improving the overall situational awareness and decision-making capabilities of the nano drone.

Another critical aspect is the development of algorithms for autonomous anomaly detection. Nano drones can be programmed to identify unusual events or patterns in the collected data. This capability is vital for security applications, environmental monitoring, and predictive maintenance. Machine learning algorithms are being trained to detect subtle anomalies that might otherwise go unnoticed, improving early warning capabilities. This autonomous anomaly detection is transforming the role of nano drones from passive data collectors to active intelligence-gathering systems.

Finally, the ethical implications of these enhanced sensor technologies and data processing capabilities must be carefully considered. The ability to collect and process vast amounts of data raises concerns about privacy and potential misuse. The development of robust security protocols and ethical guidelines is crucial to ensure responsible use of this technology. Ongoing discussions and collaboration between researchers, policymakers, and ethicists are essential for navigating the complexities of these rapidly evolving technologies and ensuring their beneficial application. The development of secure, ethical data handling protocols is not merely an afterthought but a crucial component of integrating nano drone technology into a responsible and sustainable technological framework. This ensures its benefits are realized while mitigating potential risks.

The relentless miniaturization and enhancement of nano drone capabilities inevitably trigger a counter-response, initiating a technological arms race with profound implications. The development of increasingly sophisticated nano drones, equipped with advanced sensors, AI-powered decision-making, and resilient communication systems, necessitates the parallel evolution of countermeasures designed to neutralize their threat. This dynamic interplay between offensive and defensive technologies creates a continuous cycle of innovation, pushing the boundaries of what's possible in both drone technology and counter-drone systems.

One of the primary challenges in this arms race is the inherent speed of technological advancement. The rapid pace of innovation in microelectronics, materials science, and artificial intelligence allows for frequent breakthroughs in nano drone capabilities. What constitutes a significant technological advantage today may become obsolete tomorrow. This rapid evolution necessitates continuous investment in research and development, forcing both sides to constantly adapt and innovate to maintain their technological edge. This constant pressure fosters a highly dynamic environment where breakthroughs can shift the balance of power unexpectedly. A seemingly insurmountable advantage in countermeasures today could be rendered obsolete by a novel drone design or algorithmic improvement tomorrow.

The development of effective countermeasures presents unique challenges. Traditional methods of anti-drone warfare, such as jamming, often prove ineffective against nano drones due to their small size and ability to operate in swarms. Jamming signals, for instance, might disrupt communications for larger drones, but the small size and decentralized nature of nano swarms make them more resilient to such interference. Similarly, physical interception techniques, like nets or shotguns, are problematic when dealing with swarms, often proving less effective when facing a large number of these miniature devices.

The development of countermeasures requires a multifaceted approach, combining several strategies for effective mitigation. This includes advanced sensor systems that can detect the presence of nano drones even in challenging environments. These sensors may need to operate across multiple spectrums, including acoustic, electromagnetic, and infrared, to ensure comprehensive detection capabilities. For example, acoustic sensors can detect the faint sound of drone propellers, while infrared sensors can detect the heat signature of the drone's engine. The integration of these diverse sensor modalities is crucial for comprehensive situational awareness.

Once detected, the next challenge is neutralization. Directed energy weapons, such as lasers or microwaves, show promise in disabling nano drones. However, precise targeting and power management are crucial, as excessive energy could cause collateral damage. The development of highly targeted and precisely controlled directed energy weapons is a significant area of ongoing research. Another promising area is the development of AI-powered countermeasures, using sophisticated algorithms to anticipate and counter drone manoeuvres. These systems could track the drone swarms and predict their trajectory, allowing for pre-emptive disabling measures.

The cost-effectiveness of countermeasures is a crucial factor in the ongoing arms race. Developing and deploying sophisticated countermeasures can be expensive, potentially limiting their widespread adoption. This economic aspect drives the need for innovative and cost-effective solutions, encouraging research into cheaper, more scalable alternatives. This economic constraint emphasizes the need for intelligent deployment strategies, focusing resources on high-risk areas and leveraging collaborative efforts to maximize the impact of limited resources.

Beyond the technical challenges, the societal and security implications of this technological arms race must be carefully considered. The proliferation of nano drones, coupled with the development of sophisticated countermeasures, could lead to an escalation of conflicts and create new security risks. The potential for misuse in civilian settings, such as surveillance or harassment, is also a major concern. This necessitates a proactive approach to regulate the development and deployment of these technologies, ensuring that their benefits are maximized while their risks are effectively mitigated. Ethical considerations must be central to the design, development, and deployment of both nano drones and countermeasures.

The challenge of maintaining a technological advantage in this ever-evolving landscape necessitates international cooperation. Sharing best practices, conducting joint research, and establishing international standards for the development and deployment of nano drone technology and countermeasures can help to ensure responsible innovation and prevent a dangerous escalation. Such cooperation would facilitate the development of common frameworks for ethical considerations, regulations, and oversight of this transformative technology.

Unexpected technological breakthroughs can dramatically alter the balance of power in this arms race. The discovery of a novel material, a significant advancement in AI, or an entirely new approach to drone design or countermeasures could quickly render existing systems obsolete. This unpredictability underscores the importance of continuous investment in research and development, maintaining a flexible approach that can adapt to unexpected shifts in the technological landscape.

Furthermore, the proliferation of nano drone technology raises significant concerns about accessibility and potential misuse. The relative ease of manufacturing and deploying nano drones, coupled with their small size and potential for mass production, makes them particularly susceptible to falling into the wrong hands. This concern necessitates a robust framework for controlling the distribution and use of nano drone technology, working to prevent their proliferation amongst non-state actors or those with malicious intent.

Bibliography

Antoine Bousquet, The Eye of War: Military Perception from the Telescope to the Drone (University of Minnesota Press, 2018)

Austin, R., 2010. Unmanned aircraft systems: UAVS design, development and deployment. Chichester: Wiley.

Barnhart, R.K., Hottman, S.B. and Marshall, D.M., eds., 2021. Introduction to unmanned aircraft systems. 4th ed. Boca Raton: CRC Press.

Benjamin, M., 2013. Drone warfare: Killing by remote control. New York: OR Books.

Bergen, P.L. and Rothenberg, D., eds., 2015. Drone wars: Transforming conflict, law, and policy. Cambridge: Cambridge University Press.

Boyle, M.J., 2015. The costs and consequences of drone warfare. London: Routledge.

Boyle, M.J., 2021. The Drone Age: How Drone Technology Will Change War and Peace. Highbridge

Cavallaro, J., Sonnenberg, S. and Knuckey, S., 2012. Living under drones: Death, injury, and trauma to civilians from US drone practices in Pakistan. Stanford: International Human Rights and Conflict Resolution Clinic.

Cavelty, M.D. and Mauer, V. (eds.) (2016) Drone Wars: Transforming Conflict, Law, and Policy. London: Routledge.

Celander, L. (2024) How Drones Fight: How Small Drones are Revolutionizing Warfare. Casemate

Chamayou, G. (2015). A Theory of the Drone. New York: The New Press.

Cai, G., Chen, B. and Lee, T.H., 2011. Unmanned rotorcraft systems. London: Springer.

Cao, G. and Wang, Y., eds., 2011. Nanostructures and nonmaterial: synthesis, properties, and applications. 2nd ed. Singapore: World Scientific.

Chien, S. et al., 2017. Autonomy and AI in Space and Military Systems. Pasadena: Jet Propulsion Laboratory.

Choi, J. and Rogers, J.A., 2016. Soft nanomaterials in micro/ nanosystems. New York: Springer.

Cummings, M.L., 2017. Artificial Intelligence and the Future of Warfare: Psychological and Ethical Implications. Cambridge: MIT Press.

De Lorenzo, D. and O'Neil, J. (2018) Introduction to Signals Intelligence (SIGINT). New York: Artech House.

Derek Gregory, The Everywhere War (University of California Press, 2014)

Enemark, C., 2014. Armed drones and the ethics of war: Military virtue in a post-heroic age. London: Routledge.

Ferro, F., 2020. Drone wars: Transforming conflict, law, and policy. London: Routledge.

Finn, R.L. and Wright, D., eds., 2012. Privacy, surveillance and public trust. London: Palgrave Macmillan.

Galliott, J., 2015. Military ethics and emerging technologies. London: Routledge.

George, S. and Taylor, P. (2021) Unmanned Aerial Vehicles for Intelligence, Surveillance, and Reconnaissance. 1st edn. Boca Raton: CRC Press.

Gertler, J., 2012. U.S. unmanned aerial systems. Washington, DC: Congressional Research Service.

Goss, S., 2021. Drones and the Future of Air Warfare: The Evolution of Remotely Piloted Aircraft. London: Routledge.

Grégoire Chamayou, A Theory of the Drone (The New Press, 2015)

Gregory, D., 2014. The everywhere war. Berkeley: University of California Press.

Gupta, R.C., 2020. Nano Drones: Concepts, Technology and Applications. New Delhi: Infinity Science Press.

Gusterson, H., 2016. Drone: Remote control warfare. Cambridge, MA: MIT Press.

Harrel, J.S., 2024. The Russian-Ukrainian War, 2023: A Second Year of Hell and the Dawn of Drone Warfare. Pen & Sword Military

He, J. & Xie, C. (2021) ������� [UAV Structural Design]. Beijing: Beihang University Press

IGI Global (2022) Unmanned Aerial Vehicles and Multidisciplinary Applications Using AI Techniques. IGI Global, Hershey.

Kaag, J. and Kreps, S., 2014. Drone warfare. Cambridge: Polity Press.

Kaplan, F., 2013. The rise of the machines: How computers have changed the world. London: Oneworld Publications.

Kreps, S.E Kreps, S.E. (2023) Understanding War and Peace, Cambridge University Press.., 2016. Drones: What everyone needs to know. Oxford: Oxford University Press.

Kumar, P. and Kaur, H., 2021. Nanotechnology in Aerospace and Defense. Cambridge: Cambridge

Kumar, P. and Mallick, P.K., 2018. Drone technology: principles and applications. New Delhi: IK International Publishing House.

Li, J. (2019) Unmanned Aerial Vehicle Systems: Principles and Applications. Beijing: National Defence Industry Press.

Lin, P., Abney, K. and Bekey, G.A., 2011. Robot ethics: The ethical and social implications of robotics. Cambridge, MA: MIT Press.

Lushenko, P.L. & Raman, S. (2024) The Legitimacy of Drone Warfare: Evaluating Public Perceptions. Routledge.

McKenna, J., 2019. Drone Technology in Architecture, Engineering and Construction. London: Routledge.

MDPI Books (2022) Recent Progress in UAV-AI Remote Sensing. MDPI Books, Basel.

Mulgaonkar, Y., Kumar, V. and Michael, N., 2019. Design and Control of Autonomous Quadrotors with Application to Nano-Drones. New York: Springer Briefs in Electrical and Computer Engineering.

O'Connell, M.E., 2012. What is war? An investigation in the wake of 9/11. Leiden: Martinus Nijhoff Publishers.

Pierce, M. (2017) Advanced UAV Systems for Electronic Surveillance. Boston: Artech House.

P.W. Singer, Wired for War: The Robotics Revolution and Conflict in the 21st Century (Penguin Press, 2009)

Rogers, P., 2010. Losing control: Global security in the twenty-first century. 3rd ed. London: Pluto Press.

Rogers, P., 2013. Drone warfare: Killing by remote control. Oxford: Oxford Research Group.

Rogers, J.P. (ed.), 2024. De Gruyter Handbook of Drone Warfare. De Gruyter

Rothenberg, M.A., 2013. Military tactics: A very short introduction. Oxford: Oxford University Press.

Routledge (2023) AI, Cybersecurity and Data Science for Drone and Unmanned Aerial Vehicles. Routledge, Abingdon.

Singer, P.W., 2009. Wired for war: The robotics revolution and conflict in the 21st century. New York: Penguin Press.

Springer (2022) Computational Intelligence for Unmanned Aerial Vehicles. Springer, Cham.

Strawser, B.J., ed., 2013. Killing by remote control: The ethics of an unmanned military. Oxford: Oxford University Press.

Sutton, R., 2017. Cooperative and Networked Control of Multi-Agent Systems: UAV Applications. Cham: Springer.

Valavanis, K.P. and Vachtsevanos, G.J., 2015. Handbook of Unmanned Aerial Vehicles. Dordrecht: Springer.

Yu, M. & Si, W. (2024) ���������� [UAV Flight Control Technology]. Xi'an: Northwestern Polytechnical University Press

Zhang, Q., 2017. Nano-aerial vehicles: The future of micro drones. New York: Springer.

Zhou, J., Feng, Y., Wang, Y., Wang, Z. & Zhang, G. (2021) ����� [Introduction to Unmanned Aerial Vehicles]. Beijing: Tsinghua University Press

Zhyrokhov, M. (2024) War of Drones: Drones in the Russian–Ukrainian War. 222 pp., in Ukrainian. AirHub Store, Kyiv.

Wang, L. (2021) Drone Warfare and Artificial Intelligence. Translated from Chinese by Zhang, Y. Shanghai: Science Press.

Wall, T., 2016. Law, ethics and the war on terror. London: Routledge.

Wilkinson, G. (2019) Drones and Electronic Warfare: A SIGINT Perspective. London: Springer.

Wittes, B. and Blum, G., 2015. The future of violence: Robots and germs, hackers and drones. New York: Basic Books.

Zenko, M., 2016. Drone warfare: The future of conflict. New York: Council on Foreign Relations Press.

About the Author

The Author is trained lawyer, President of an Educational Foundation,Representative of Brazilian institutions in Europe either involvedin military subjects (ADESG) either commercial and industrial- (the Federation of Chambers of Commerce and Industry ofSouth America). The Author also it is a researcher in geopolitics andnational defence. He is actually is involved in a number of professionalactivities of several areas like international partnerships havingin common development and cooperation. Author of several booksbooks, essays and articles concerning its activity and passed professionalexperience. He give universities lectures, conferences, and expertiseassessments.Read more at https://observare.autonoma.pt/investigador/artur-victoria/.

Read more at https://observare.autonoma.pt/investigador/artur-victoria/.

www.ingramcontent.com/pod-product-compliance
Lightning Source LLC
LaVergne TN
LVHW090305210926
848480LV00001B/136
* 9 7 9 8 2 3 1 6 3 0 5 3 0 *